Algorithms and Model Formulations
in Mathematical Programming

NATO ASI Series

Advanced Science Institutes Series

A series presenting the results of activities sponsored by the NATO Science Committee, which aims at the dissemination of advanced scientific and technological knowledge, with a view to strengthening links between scientific communities.

The Series is published by an international board of publishers in conjunction with the NATO Scientific Affairs Division

A **Life Sciences**	Plenum Publishing Corporation
B **Physics**	London and New York
C **Mathematical and**	Kluwer Academic Publishers
Physical Sciences	Dordrecht, Boston and London
D **Behavioural and**	
Social Sciences	
E **Applied Sciences**	
F **Computer and**	Springer-Verlag
Systems Sciences	Berlin Heidelberg New York
G **Ecological Sciences**	London Paris Tokyo
H **Cell Biology**	

Algorithms and Model Formulations in Mathematical Programming

Edited by

Stein W. Wallace

Chr. Michelsen Institute
Fantoftvegen 38, N-5036 Fantoft, Norway

Springer-Verlag
Berlin Heidelberg New York London Paris Tokyo
Published in cooperation with NATO Scientific Affairs Division

Proceedings of the NATO Advanced Research Workshop on Algorithms and Model Formulations in Mathematical Programming held in Bergen, Norway, June 15–19, 1987.

ISBN-13: 978-3-642-83726-5   e-ISBN-13: 978-3-642-83724-1
DOI: 10.1007/978-3-642-83724-1

Library of Congress Cataloging-in-Publication Data. NATO Advanced Research Workshop Algorithms and Model Formulations in Mathematical Programming (1987: Bergen, Norway) Algorithms and model formulations in mathematical programming / edited by Stein W. Wallace. p. cm.—(NATO ASI series. Series F, Computer and systems sciences; vol. 51) "Proceedings of the NATO Advanced Research Workshop Algorithms and Model Formulations in Mathematical Programming, held in Bergen, Norway, June 15–19, 1987"—T.p. verso. "Published in cooperation with NATO Scientific Affairs Division."

1. Programming (Mathematics)—Congresses. 2. Mathematical models—Congresses. I. Wallace, Stein W., 1956-. II. North Atlantic Treaty Organization. Scientific Affairs Division. III. Title. IV. Series: NATO ASI series. Series F, Computer and system sciences; vol. 51. T57.7.N36 1987
519.7—dc 19 89-4264

© Springer-Verlag Berlin Heidelberg 1989
Softcover reprint of the hardcover 1st edition 1989
Printing: Druckhaus Beltz, Hemsbach; Binding: J. Schäffer GmbH & Co. KG, Grünstadt
2145/3140-543210 – Printed on acid-free paper

PREFACE

The NATO Advanced Research Workshop (ARW) "Algorithms and Model Formulations in Mathematical Programming" was held at Chr. Michelsen Institute in Bergen, Norway, from June 15 to June 19, 1987. The ARW was organized on behalf of the Committee on Algorithms (COAL) of the Mathematical Programming Society (MPS). Co-directors were Jan Telgen (Van Dien+Co Organisatie, Utrecht, The Netherlands) and Roger J-B Wets (The University of California at Davis, USA). 43 participants from 11 countries attended the ARW.

The workshop was organized such that each day started with a 90-minute keynote presentation, followed by a 45-minute plenary discussion. The first part of this book contains the contributions of the five keynote speakers. The plenary discussions were taped, and the transcripts given to the keynote speakers. They have treated the transcripts differently, some by working the discussions into their papers, others by adding a section which sums up the discussions. The plenary discussions were very interesting and stimulating due to active participation of the audience.

The five keynote speakers were asked to view the topic of the workshop, the interaction between algorithms and model formulations, from different perspectives. On the first day of the workshop Professor Alexander H.G. Rinnooy Kan (Erasmus University, Rotterdam, The Netherlands) put the theme into a larger context by his talk "Mathematical programming as an intellectual activity". This is an article of importance to any mathematical programmer who is interested in his field's history and present state.

Professor Walter Murray (Stanford University, California, USA) views the topic from the standpoint of linear programming. In his talk "Methods for linear programming", he discussed, among other things, the relation between the simplex method and versions of the barrier method, such as for example Karmarkar's projective algorithm.

There exists a certain kind of "competition" between mathematical programming and scenario analysis in simulation. Professor Roger J-B Wets (University of California at Davis, USA) addresses this competition in his paper, by showing how the two can be combined into a powerful tool. His approach is particularly useful for problems where deterministic function evaluations (i.e. simulations) represent the only possible way of attacking a problem. He demonstrates how one, by closely examining which simulations one performs, can "simulate" a multistage stochastic programming approach.

Professor John Mulvey (Princeton University, New Jersey, USA) considers network models with nonlinear objectives. He describes many interesting and genuine applications from economics, engineering and management. In addition he discusses solution procedures for the resulting large-scale problems.

Dr. Ellis Johnson (IBM - Yorktown Heights, New York, USA) views mixed integer programming problems from two points of view: model generation in an understandable format and formulation or reformulation such that the model can be solved. The focus is on preprocessing, constraint generation, and column generation.

The second part of this book contains extended abstracts for all presentations made at the workshop. A list of addresses is also provided, and interested readers are encouraged to use this list if they wish more details about the topics of the extended abstracts.

The organizers of this ARW are deeply indebted to The NATO Science Affairs Division for the financial support. Other sponsors were Chr. Michelsen Institute, The Royal Norwegian Council for Scientific and Industrial Research, The University of Bergen and Bergen Scientific Center (IBM).

The organizing committee wishes to thank Sverre Storøy for his efforts with the social events, and Ragnhild Engelsen (now Nordanger) for her enthusiasm and long hours to make sure the organizational aspects work smoothly. Other members of the local committee were Åsa Hallefjord, Kurt Jørnsten, Ronny Aboudi, Reidun Helming and Anne Pettersen. Without their efforts the workshop would not have taken place.

Chr. Michelsen Institute, November 1988   Stein W. Wallace

TABLE OF CONTENTS

Modeling and Strong Linear Programs for Mixed Integer Programming

Ellis L. Johnson
Department of Mathematical Sciences
IBM Research Center
T.J. Watson Research Center
P.O. Box 218
Yorktown Heights, NY 10598

Abstract: Mixed integer programming modeling is considered from two points of view: getting the model correctly generated in an understandable form, and formulating or reformulating the model so that the problem can be solved. For the former considerations, a relational approach is presented. For the latter, three techniques are discussed: preprocessing, constraint generation, and column generation. For all three techniques, mixed integer problems are considered. For column generation, two problem classes (cutting stock and crew scheduling) for which column generation techniques are classical, are presented in a unified framework and then clustering problems are discussed in the same framework. In the constraint generation section, some constraints based on mixed 0-1 implication graphs are presented.

NATO ASI Series, Vol. F51
Algorithms and Model Formulations
in Mathematical Programming
Edited by Stein W. Wallace
© Springer-Verlag Berlin Heidelberg 1989

1. Introduction.

In modeling mixed integer programs, it is important to distinguish those aspects involving correctly capturing the logic of the constraints of the problem and those considerations involving improving the model so that it can be solved once it was formulated. Section 2 is concerned mainly with the former whereas the rest of the paper is mainly concerned with the latter.

There are two main reasons that integer programming models are not used more in practice. One is that after much work to get the model set up and the data collected, codes may not be able to solve or even get reasonably good solutions to problems that are small enough that the linear programming problems are easy. The other is the difficulty in getting the model set up, the data collected in the form needed, and the solution understood. This latter difficulty is also true for linear programming. For linear programs, at least one can reasonably expect that codes can be found to solve the linear programs once they have been put together. It is a reasonable contention that for linear programs, the modelling difficulty is the main one hindering wider use. In other words, computational difficulty of linear programs is not a major hindrance to wider use. Matrix generation, even intelligently done, can frequently take the same order of time as solving the problem. Furthermore, once a model gets into production making changes in the model to capture changes in the real problem can be difficult. Time of programmers and applications specialist is probably more of a consideration than computer time.

While it is, of course, true that some users have difficult linear programs, and competitive pressure exists to have better performance, the claim is that even major reductions in linear programming solution times would not cause a flood of new users. The converse claim is made that if modeling were just collecting data, matrix generation was non-existent, and solutions were immediately comprehensible, then there would be a large increase in the number of users of linear programming codes. How to make it so easy is not known but is something to work toward.

In integer programming, computer running times can be excessive and are, by comparison with linear programs, a major consideration. The main approach to improving running times is to improve the linear programming relaxation. We discuss two techniques here. One is constraint generation based on logical considerations [4] . One possible type of constraint that has proven effective [2] in fixed charge problems is detailed in section 4.

Sections 5 - 7 give a column generation approach that could be considered as part of the modelling process or as an automatic sort of decomposition. The emphasis here is in improving the linear programming relaxation by solving for integer solutions to more easily solved subproblems. The two classical examples are the cutting stock and the crew scheduling problems.

2. Modeling Tools

To clarify terminology, we refer to "the problem" as being the user's problem. For example, in a power generation problem, the problem is when to start-up and shut-down plants so as to minimize operating costs subject to having sufficient power being generated to meet demand, where demand depends on the time of the day. That is, some power plants actually exist and must be operated and somebody has to make a schedule. That is "the problem" . The problem data is somewhere, hopefully, available to the decision maker.

The "model" is here either a linear or integer program, i.e. $Ax = b, x \geq 0$ and perhaps integer, minimize cx. The model is in one sense very simple. In practice, for example in the above power generation case, it may be a thick notebook explaining the constraints, the data needed, the variables, etc. Also, a model can almost never capture all of a problem; one can only hope that the model is realistic enough to make the solution useful to the decision maker.

The "matrix" is a particular realization of a model that has actual numbers coming from the problem data. A matrix generator is a code that fills up the matrix based on the model description and the problem data. One also refers to a matrix generator as a code that first generates a code which then generates the matrix. The modeler is faced with another level of difficulty: is the solution wrong because the generator is not working correctly. Particularly for special codes written to generate the matrix, there may be a bug in the code itself. Or the user may not have correctly used it, e.g. he may have used the wrong notation. A similar problem occurs in trying to track down bad coefficients: was the problem data given just plain wrong, or was it put in the wrong place, or were the units not matched, etc. On the other hand, it may be that the answer was correct but just looked strange and, nevertheless, must be understood before it can be used in the decision making process.

Any tool that is introduced to help the modeler be comfortable is welcome. However, let me make a pitch for mathematical programmers to avoid new languages and new constructs. In other words, let us try to join the main stream of commercial data processing and avoid self induced isolation. One example of joining the main-stream is the tying of linear programming to spread-sheets. It is not to say that one should learn spread-sheet systems in order to do linear programming modelling. But if one is using spread-sheets already and is aware of the possibilities of letting linear programs be invoked automatically to optimize some of the parameters in the spread-sheet, then the system itself does the matrix generation based on data already in the spread-sheet and in the form needed. Furthermore and perhaps most importantly, the user knows what the answer means. If he wants to check the solution he can run the spread sheet with those values and it is just like the scenario analysis he was already doing except that somebody else (the linear program) gave him the values to try.

Spread-sheet systems are certainly some of the most popular codes available. As further evidence of the fact that this type of use is in the main-stream, one of these linear programming codes got a half page review in successive issues of the Sunday business section of a main New York newspaper. The reviewer said that he not even be reviewing it if it was just a nice front-end to LINDO. He was reviewing it because it was tied to and an easily usable part of a popular spread-sheet system.

My nomination for a more general part of main-stream computing to hitch mathematical programming to is relational data base. This is not a new idea [5,14], and the ideas presented here are based on work with Maria Cristina Esposito [6] and John Forrest [7]. Relational data representations are natural results of structured modelling [8], and then the ideas presented here are even more coherent. In any case, if the users problem data is already in relational tables, then matrix generation as presented here is easier. Specifically, the use of SQL as a matrix generation tool will be discussed. To clarify the terminology, SQL is sometimes used to denote a data-base system whose language is SEQUEL [1]. In common use, SQL also means the language, and we follow that usage. The commands are all commands to manipulate tables.

The example used to illustrate the concepts is a slightly simplified version of a production distribution problem with single sourcing requirement [15]. The problem is to decide how much of each product to produce at the plants, how to ship to warehouses, and tranship to demand-centers subject to the constraint that a warehouse has to ship all of the demand for all products to any demand-center that it ships to. In other words, each demand-center is assigned a single warehouse that must meet all of its demands for the several products.

A basic building block of any model is indices, or subscripts, or names. In this case, there are four index sets with what will be called index-set names: PLANT, WHSE, PRODUCT, CEN-TER. The user must provide four user-tables of user data. In describing a table, its columns are called fields and its rows are called records following relational terminology and to avoid confusion with rows and columns of matrices. The four user tables have table names **PRODUCTION, SHIPCOST, TRANSHIP,** and **DEMAND** with fields as below:

PRODUCTION: PLANT, PRODUCT, CAPACITY, COST
SHIPCOST: PLANT, WHSE, COST
TRANSHIP: WHSE, CENTER, COST
DEMAND: CENTER, PRODUCT, AMOUNT

Note that a field names in different tables can be the same, e.g. cost appears in the first three tables, but are distinguished by being **PRODUCTION.COST, SHIPCOST**.COST, and **TRANSHIP**.COST. Also, the index-set names, e.g. PLANT, are defined by the user tables, and a certain degree of consistency and checking can be done at this level: e.g. a record of the table **PRODUCTION** for which the capacity is zero can be deleted and there need be no plant present in the table **SHIPCOST** which is not present in the table **PRODUCTION**, and visa versa. Further, a warehouse need not be included unless it is possible to ship every product to it. This latter condition is due to the single-sourcing constraint.

All that has been done so far is to try to specify the data that is needed to define the problem. The model has not yet been addressed. The meaning of the user tables should be fairly self-explanatory. Note that the units for the products is assumed to be the same in the tables so that shipping costs do not depend of the product.

The model is described in [15] and we go directly to the preferred form with the strong linear program. This form can be formed using the preprocessing techniques in the next two sections. Suffice it to say that the point of [15] was that the difference in running times is incomparably better with the form presented here.

There are three sets of variables in the model: **PRODUCE, SHIP,** and **ASSIGN,** where the variables in the set **ASSIGN** are of type 0-1. In fact, let these three names be the table names for column variables having fields as below:

PRODUCE: PLANT, PRODUCE, VALUE

SHIP: PLANT, WHSE, PRODUCT, VALUE

ASSIGN: WHSE, CENTER, VALUE

In every case, VALUE is the solution value to be filled in by solving the mixed integer program. In terms of usual variable names, if a plant at topeka produces a product nachos, then there is a variable named produce-topeka-nachos. That is, a record of the table is a variable, and its name is all but the last field which is reserved for the solution value.

Similarly there are three types of rows represented by tables with names: **PRODROW, SHIPROW,** and **CENTROW.** Their fields are:

PRODROW: PLANT, PRODUCT, VALUE

SHIPROW: WHSE, PRODUCT, VALUE

CENTROW: CENTER, VALUE

So far, we have four user tables and six variable tables, of which three are column variables and three are row variables. To specify how the variable tables, ignoring the fields VALUE that are, for now, dummy fields to be filled in at solution time, are formed in this example from the user tables, the concepts of a table select and table join need to be explained. A table select means to form a new table as a sub-table of a given table where the new table has a subset of the fields and subset of the records of the given table. An example has already been discussed when it was pointed out that the table **PRODUCTION** need not contain any record for which the field CAPACITY is zero; that is, the table **PRODUCTION** could be replaced by a the table formed by deleting those records for which CAPACITY is zero. Another more illuminating example is the table **PRODUCE**: it is a subtable of **PRODUCTION** where the fields PLANT and PRODUCT have been selected. Similarly, **ASSIGN** is the subtable of **TRANSHIP** with fields WHSE and CEN-TER. The table **PRODROW** is identical, except for VALUE, to **PRODUCE**. The table **CENTROW** is the subtable of **DEMAND** with field CENTER except that duplicate records must be deleted.

That only leaves the tables **SHIP** and **SHIPROW**. Here we need a table join. To join two tables in simple form means to form a new table whose fields are the union of the fields of the two tables and whose records are the "outer product" of the records of the two tables. That is, for each pair of records, one from each table, a record of the new table is formed by simply concatenating the two records. A predicated join is specified by identifying pairs of fields of the two tables as being the same so that those two fields appear as only one field of the new table and a record of the new table appears for every pair of records of the two tables having the same entries in the identified pairs of fields.

The table **SHIP** is the predicated join, with identified fields **PRODUCTION.PLANT** and **SHIPCOST.PLANT**, of the subtable of production having fields PLANT and PRODUCT and the subtable of **SHIPCOST** having fields PLANT and WHSE. In this way, **SHIP** has records only for PLANT, PRODUCT, WHSE triples for which the plant produces the product and the plant can ship to the warehouse. The table **SHIPROW** is the subtable of **SHIP** having fields WHSE and PRODUCT.

We give a small example. Let the four user tables be:

PRODUCTION					DEMAND		
PLANT	PRODUCT	CAPACITY	COST		CENTER	PRODUCT	AMOUNT
topeka	chips	200	230		east	chips	200
topeka	nachos	800	280		east	nachos	50
ny	chips	600	255		south	chips	250
					south	nachos	180
					west	chips	150
					west	nachos	300

SHIPCOST				TRANSHIP		
PLANT	WHSE	COST		WHSE	CENTER	COST
topeka	topeka	1		topeka	east	60
topeka	ny	45		topeka	south	30
ny	ny	2		topeka	west	40
ny	topeka	45		ny	east	10
				ny	south	30
				ny	west	80

Leaving off the field VALUE, the three column variable tables **PRODUCE, SHIP** and **ASSIGN** are:

PRODUCE		SHIP			ASSIGN	
PLANT	PRODUCT	PLANT	WHSE	PRODUCT	WHSE	CENTER
topeka	chips	topeka	topeka	chips	topeka	east
topeka	nachos	topeka	topeka	nachos	topeka	south
ny	chips	topeka	ny	chips	topeka	west
		topeka	ny	nachos	ny	east
		ny	ny	chips	ny	south
		ny	topeka	chips	ny	west

The three row variable tables **PRODROW, SHIPROW,** and **CENTROW** are:

PRODROW		SHIPROW		CENTROW
PLANT	PRODUCT	WHSE	PRODUCT	CENTER
topeka	chips	topeka	chips	east
topeka	nachos	topeka	nachos	south
ny	chips	ny	chips	west
		ny	nachos	

So far only the variables have been defined. To form the matrix, more table joins will be needed. The matrix will be formed in blocks, i.e. entries in a part of the matrix corresponding to a row table and a column table. First a short-hand specification will be given. It is in the nature of a matrix generation language but can actually be used to generate SQL statements to form the required tables.

BLOCK NAME	ROW NAME	COLUMN NAME	ENTRY
BLOCK11	PRODROW	PRODUCE	-1
BLOCK12	PRODROW	SHIP	1
BLOCK22	SHIPROW	SHIP	-1
BLOCK23	SHIPROW	ASSIGN	**DEMAND.AMOUNT**
BLOCKCAP	UPPER	PRODUCE	**PRODUCTION.CAPACITY**
BLOCKRHS3	CENTROW	UPPER,LOWER	1
BLOCKOBJ1	OBJECTIVE	PRODUCE	**PRODUCTION.COST**
BLOCKOBJ2	OBJECTIVE	SHIP	**SHIPCOST.COST**
BLOCKOBJ3	OBJECTIVE	ASSIGN	+ /(PRODUCT)
			DEMAND.AMOUNT
			x**TRANSHIP.COST**

The matrix is partitioned into blocks, where each block has rows and columns that are either names of row variable tables and column variable tables or are key words such as "upper", or "objective". The tables for the blocks of the matrix are explicitly given below for this example:

BLOCK11

PLANT	PRODUCT	COEF
topeka	chips	-1
topeka	nachos	-1
ny	chips	-1

BLOCK12

PLANT	PRODUCT	WHSE	COEF
topeka	chips	topeka	1
topeka	chips	ny	1
topeka	nachos	topeka	1
topeka	nachos	ny	1
ny	chips	ny	1
ny	chips	topeka	1

BLOCK22

WHSE	PRODUCT	PLANT	COEF
topeka	chips	topeka	-1
topeka	chips	ny	-1
topeka	nachos	topeka	-1
topeka	nachos	ny	-1
ny	chips	ny	-1
ny	chips	topeka	-1
ny	nachos	topeka	-1

BLOCK23

WHSE	PRODUCT	CENTER	COEF
topeka	chips	east	200
topeka	chips	south	250
topeka	chips	west	150
topeka	nachos	east	50
topeka	nachos	south	180
topeka	nachos	west	300
ny	chips	east	200
ny	chips	south	250
ny	chips	west	150
ny	nachos	east	50
ny	nachos	south	180
ny	nachos	west	300

BLOCKCAP		
PLANT	PRODUCT	COEF
topeka	chips	200
topeka	nachos	800
ny	chips	600

BLOCKRHS3	
CENTER	COEF
east	1
south	1
west	1

BLOCKOBJ1		
PLANT	PRODUCT	COEF
topeka	chips	230
topeka	nachos	280
ny	chips	255

BLOCKOBJ2			
PLANT	WHSE	PRODUCT	COEF
topeka	topeka	chips	1
topeka	topeka	nachos	1
topeka	ny	chips	45
topeka	ny	nachos	45
ny	ny	chips	2
ny	topeka	chips	45

BLOCKOBJ3		
WHSE	CENTER	COEF
topeka	east	60x(200 + 50)
topeka	south	30x(250 + 180)
topeka	west	40x(150 + 300)
ny	east	10x(200 + 50)
ny	south	30x(250 + 180)
ny	west	80x(150 + 300)

In the short-hand, "upper" or "lower" is a key-word that, depending on its location means either upper or lower bounds on the column variables or on the values of the right-hand side. The defaults taken here are that column variables have upper bound infinity unless it is integer in which case the upper bound is 1. All other bounds are defaulted at 0, so the default for rows is that they are of type = 0. The notation "upper,lower" means both bounds are equal to the specified value. This is the way that equations are specified.

The coefficient matrix in full matrix form is given below. The abbreviation there are: n = ny, t = topeka, c = chips, n = nachos, e = east, s = south, and w = west. The row denoted "up" is an upper bound row on the variables and comes from the block table **BLOCKCAP** and from the default of upper bound equals one for the integer variables **ASSIGN**. The objective row denoted "obj" comes from the three tables **BLOCKOBJ1, BLOCKOBJ2, BLOCKOBJ3**.

	PRODUCE			SHIP						ASSIGN						
	t c	t n	n c	tt c	tt n	tn c	tn n	nn c	nt c	t e	t s	t w	n e	n s	n w	
	-1			1		1										= 0
		-1			1		1									= 0
			-1					1	1							= 0
				-1					-1	200	250	150				= 0
					-1					50	180	300				= 0
						-1		-1					200	250	150	= 0
							-1						50	180	300	= 0
										1			1			= 1
											1			1		= 1
												1			1	= 1
up	200	800	600							1	1	1	1	1	1	
obj	230	280	255	1	1	45	45	2	45	15K	9.9K	18K	2.5K	9.9K	36K	

Returning to the short-hand notation for a block of the matrix, if the row table and the column table have fields with the same names (remember that the modeler has created these tables), then the default is to join the row table and column table using a predicate join identifying the pairs of fields with the same names. When it comes to filling in the coefficients, a third table (which may be a scalar such as 1 or -1) must be specified and its fields must be a subset of the union of the fields for the row table and the column table. In the example above, **BLOCK23** has row table **SHIPROW** and column table **ASSIGN**. Those tables and their predicated join are below:

SHIPROW		ASSIGN		JOIN		
WHSE	PRODUCT	WHSE	CENTER	WHSE	PRODUCT	CENTER
topeka	chips	topeka	east	topeka	chips	east
topeka	nachos	topeka	south	topeka	chips	south
ny	chips	topeka	west	topeka	chips	west
ny	nachos	ny	east	topeka	nachos	east
		ny	south	topeka	nachos	south
		ny	west	topeka	nachos	west
				ny	chips	east
				ny	chips	south
				ny	chips	west
				ny	nachos	east
				ny	nachos	south
				ny	nachos	west

The table **DEMAND** has fields: CENTER, PRODUCT, AMOUNT. To get **BLOCK23**, a join is done with the two tables join (above) and **DEMAND**, but the field AMOUNT gets renamed COEF in **BLOCK23**.

A more complicated case is to get the objective function coefficients of **ASSIGN**. Some notation is there in the short-hand, but what needs to be done is the following. First, form the table **TOTAL** by adding all of the AMOUNT in the table **DEMAND** for a fixed CENTER and different PRODUCT:

TOTAL

CENTER	AMOUNT
east	250
south	430
west	450

Doing a join of the table total and the table tranship gives

TEMPTABLE

WHSE	CENTER	COST	AMOUNT
topeka	east	60	250
topeka	south	30	430
topeka	west	40	450
ny	east	10	250
ny	south	30	430
ny	west	80	450

The coefficient in the objective function for assign is the product of the two fields cost and amount in the above table.

Once we have a translation of the short-hand into an executable code, we have a data-independent model representation. All that is required for this model is four user tables having fields as specified.

There are several advantages of this approach to matrix generation. Gathering the appropriate data can be a difficult task in modeling. Here, the matrix generator stays in a data-base environment. The matrix itself is a table. And, every column and row table has a field VALUE, but that field can also be transferred to the users tables, e.g. **PRODUCE**.VALUE could go into a new field **PRODUCTION**.HOWMUCH of the users table. Then production quantities are right beside capacities and costs in that table. Questions such as allowing long names of variables disappear since the names are concatenations of names in fields of a table, and how big each can be is a data-base question. How many fields there can be is, as well, a data base question. When the linear programming optimizer comes to solving the problem, it needs to convert names to indices anyway, so names are purely a modelling and solution analysis aid.

An advantage coming from this way of matrix generation is that the coefficients are closer to the user data, in fact they may be views of the user data. That is, tables need not actually be formed but can be specified as a subtable of other tables. In this way, questions such as data consistency or meaning are more easily addressed.

Another and perhaps bigger advantage is that the model structure is retained in the matrix representation. Once the model is specified and understood, data from the user can be checked even in the users table, e.g. plants should appear in the table shipcost or else the production there can never go anywhere. Furthermore, preprocessing can be done on a block scale. For example, non-negativity of the variables **PRODUCE** is redundant because they have a negative identity block with equality constraints and only positive coefficients in the other rows. This information can be seen by only looking at the short-hand description of the blocks. In this case, there are upper bounds on those variables so they cannot be declared free, but in other cases free variables might be identified at the block level. In general, the model structure can be more easily seen than from a matrix representation of the problem, and a picture of the block structure of the total matrix can be useful in gaining an understanding of a model.

The block representation makes it easier to build subproblems or super problems of given problems, since the matrix is just a union of various blocks. Furthermore, the frame work includes current MPS input format in which there is only one block with three fields: column-name, row-name, and coefficient. More interestingly, any block whose non-zero entries are too complicated can be left to the user to make up a table that is a sparse matrix representation of that block. Other blocks that are straight-forward can be done in a simpler way, as illustrated.

One last remark is that subsets of index-sets can be useful. An example is in problems with time periods, where it can be useful to have the time periods up to a given time or the time period preceding a given time period. Allowing a block to be specified more than once is also useful. Multiple specifications of blocks can be handled by adding together the entries (with a warning of its occurrence). An example is in a time period problem where a block representing transfer from one time period to the next may have a main diagonal of -1 and a subdiagonal of + 1. This block can conveniently be specified by specifying the two parts separately. In this case, numbers are not actually added together because the non-zeroes do not overlap. Another variant is that we may want separate names for the same index set. An example is, again, time periods. A column variable may be indexed by time twice, e.g. time investment was made and time of return. Such a column variable may appear in a block whose row variables have time only once. In this case, we need to know which fields to identify, if any, in doing the join to form the block. That is to say, for some blocks the fields should not be identified in doing the predicate join and in others, as in all of the blocks in the example here, they should be identified, depending on the model. If none of the fields are identified in forming some block, the result is a dense matrix except that the table of non-zero entries need not be full so that the result can be said to be any sparse matrix depending on the coefficient table given.

3. MIP Preprocessing

Preprocessing integer programming problems in order to improve the linear programming approximation to the integer problem and thereby improve the performance of branch-and-bound codes can be of critical practical importance. Even for linear programs, preprocessing or reducing the problem can cause significant improvements in performance [18]. In this section, a symmetric primal-dual procedure for linear and mixed integer programming problems is given. The notion of implication graphs is extended from the pure 0-1 case, and their use is discussed.

Consider a linear program in the symmetric form:

$$\ell \leq x \leq u,\, s \geq 0,\, t \geq 0,$$
$$Ax - Is + It = b,$$
$$minimize\ z = cx - \lambda s + \mu t$$

The dual linear program is:

$$\lambda \leq \pi \leq \mu,\, \sigma \geq 0,\, \tau \geq 0,$$
$$\pi A + \sigma I - \tau I = c,$$
$$maximize\ \zeta = \pi b + \sigma \ell - \tau u$$

It is assumed that $\ell \leq u$ and $\lambda \leq \mu$ and that ℓ or λ could be $-\infty$ and u or μ could be $+\infty$ The standard form of a linear program has non-negativity restrictions on x and equality in the primal problem. This form of the problem is achieved by letting $\ell = 0, u = +\infty,$ $\lambda = -\infty,$ and $u = +\infty$. In the form we have written the problems, both the primal and dual problems are always feasible; however, we nevertheless refer to the primal problem as <u>infeasible</u> if $z = +\infty$ for all solutions x, s, t, and we refer to the dual problem as being <u>infeasible</u> if $\zeta = -\infty$ for all dual solutions π, σ, τ. Also, when we refer to the <u>feasible region</u> for either problem, we mean the set of solutions such that the objective function is finite.

In a mixed integer program, some of the primal variables x are required, in addition, to be integer valued. Let J_I denote the indices of such integer variables. Differences in the procedures due to the integrality restrictions, x_j integer for $j \in J_I$, will be discussed.

First, a procedure for determining tighter implied bounds x and π. Let ℓ, u and λ, μ denote the given bounds and let ℓ', u' and λ', μ' denote tighter implied bounds, i.e. $\ell' \leq x \leq u'$ is implied by feasibility and optimality:

$$\ell \leq x \leq u,\, s \geq 0,\, t \geq 0,$$
$$Ax - Is + It = b,$$
$$z^* = cx - \lambda s + \mu t,\ where\ z^*\ is\ the\ optimum\ value\ of\ z,\ and$$
$$s_I = 0\ whenever\ \lambda_I = -\infty\ and\ t_I = 0\ whenever\ \mu_I = +\infty.$$

Denote $P_i = \{j \mid a_{ij} > 0\}$ and $N_i = \{j \mid a_{ij} < 0\}$. Let

$$m_i = \sum_{j \in P_i} a_{ij} \ell'_j + \sum_{j \in N_i} a_{ij} u'_j , \quad \text{and}$$

$$M_i = \sum_{j \in P_i} a_{ij} u'_j + \sum_{j \in N_i} a_{ij} \ell'_j$$

Thus, m_i is a lower bound on the sum $\sum a_{ij} x_j$ subject to the bounds $\ell' \le x \le u'$. Similarly, M_i is an upper bound on the same sum.

<u>Theorem 2.1:</u> If $\mu_i = +\infty$ and $M_i < b_i$, then the primal problem is infeasible, i.e. has objective function $z = +\infty$. Similarly, the primal problem is infeasible if $\lambda_i = -\infty$ and $m_i > b_i$.

Proof: The theorem would be obvious if m_i and M_i were defined in terms of ℓ, u instead of ℓ', u'. But the tighter bounds are implied by feasibility and optimality, so the contradiction obtained is that feasibility and optimality imply an evident infeasibility.

<u>Theorem 2.2:</u> If $\mu_i = +\infty$ and $m_i > b_i$, then the ith primal constraint is inactive, and μ'_i can be lowered to λ_i, i.e. π_i can be fixed to λ_i. Similarly, if $\lambda_i = -\infty$ and $M_i < b_1$, then the ith primal constraint is inactive and λ'_i can be raised to μ_i, i.e. π_i can be fixed to μ_i.

Proof: The ith constraint being inactive means that one of the two slack variables s_i, t_i will be zero and the other positive in every optimal solution. If $\mu_i = +\infty$, then $t_i = 0$ in any feasible solution; and if a $m_i > b_i$, then s_i is positive in any optimal solution to the primal problem. By complimentary slackness, π_i is equal to λ_i in any optimal solution. Hence, μ_i can be lowered to λ_i.

In case $\mu_i = +\infty$ and $m_i > b_i$, then it is not possible that $\lambda'_i > \lambda_i$ is implied by the other constraints. The reason is that for the original constraints to imply $\pi_i > \lambda_i$ it is necessary that there be some columns j with $u_j = +\infty$ and $a_{ij} < 0$ or with $\ell_j = -\infty$ and $a_{ij} > 0$, and in either case it must be that $m_i = -\infty$ so $m_i > b_i$ is impossible.

For example, consider the primal problem with $u_j = +\infty$:

$$0 \leq x \leq +\infty, s \geq 0,$$

$$Ax - Is = b,$$

$$minimize \ z = cx,$$

and the associated dual problem:

$$0 \leq \pi \leq +\infty, \sigma \geq 0,$$

$$\pi A + \sigma I = c,$$

$$maximize \ \zeta = \pi b.$$

If some inequality is implied by $x \geq 0$, e.g.

$$x_j + x_k - s_i = b_i, \quad \text{where} \ b_i < 0,$$

then $M_i > b_i$ as in Theorem 2.2 and μ_i' can be lowered to 0. In other words $s_i > 0$ is implied by $x \geq 0$, and the optimal solution is not changed by letting t_i be present with cost $\mu_i = 0$.

Theorem 2.3: If $\lambda_i = -\infty$ and $m_i + |a_{ik}|(u_k - \ell_k) > b_i$, then ℓ_k or u_k can be changed to:

$$u_k = (b_i + a_{ik}\ell_k - m_i)/a_{ik}, \quad \text{if} \ a_{ik} > 0, \ \text{or}$$

$$\ell_k = (-b_i - a_{ik}u_k + m_i)/(-a_{ik}), \quad \text{if} \ a_{ik} < 0.$$

Similarly, if $\mu_i = \infty$ and $M_i - |a_{ik}|(u_k - \ell_k) < b_i$, then ℓ_k or u_k can be changed to:

$$\ell_k = (b_i + a_{ik}u_k - M_i)/a_{ik}, \quad \text{if} \ a_{ik} > 0, \ \text{or}$$

$$u_k = (-b_i - a_{ik}\ell_k + M_i)/(-a_{ik}), \quad \text{if} \ a_{ik} < 0.$$

Proof: Again, the theorem would be obvious if m_i and M_i were defined in terms of ℓ, u instead of ℓ', u'. As for the proof of Theorem 2.1, the tighter bounds ℓ', u' can be used because they are implied by feasibility and optimality.

When a bound can be improved, e.g. when

$$\lambda_i = -\infty \ \text{and} \ m_i + |a_{ik}|(u_k - \ell_k) > b_i \ \text{and} \ a_{ik} > 0$$

then ℓ_k can be increased as specified in theorem 2.3 to say, ℓ_k' . But, more can be implied. For example, if some $a_{ij} > 0, j \neq k$, then

$$a_{ij}(x_j - \ell_j) + a_{ik}(x_k - \ell_k) \leq b_i - m_i \text{ is valid.}$$

More generally, whenever $\lambda_i = -\infty$ and

$$m_i + |a_{ij}| (u_{ij} - \ell_{ij}) + |a_{ik}| (u_k - \ell_k) > b_i,$$

then one of the inequalities below

$$a_{ij}(x_j - \ell_j) + a_{ik}(x_k - \ell_k) \leq b_i - m_i, \text{ if } a_{ij} > 0 \text{ and } a_{ik} > 0;$$
$$- a_{ij}(u_j - x_j) + a_{ik}(x_k - \ell_k) \leq b_i - m_i, \text{ if } a_{ij} < 0 \text{ and } a_{ik} > 0;$$
$$- a_{ij}(u_j - x_j) - a_{ik}(u_k - x_k) \leq b_i - m_i, \text{ if } a_{ij} < 0 \text{ and } a_{ik} < 0.$$

is valid and is not implied by $\ell_j \leq x_j \leq u_j$ and $\ell_k \leq x_k \leq u_k$. On the other hand, if $\mu_i = +\infty$ and

$$M_i - |a_{ij}| (u_j - \ell_k) - |a_{ik}| (u_k - \ell_k) < b_i,$$

then one of the inequalities below

$$a_{ij}(u_j - x_j) + a_{ik}(u_k - x_k) \leq M_i - b_i, \text{ if } a_{ij} > 0 \text{ and } a_{ik} > 0;$$
$$- a_{ij}(x_j - \ell_j) + a_{ik}(u_k - x_k) \leq M_i - b_i, \text{ if } a_{ij} < 0 \text{ and } a_{ik} > 0;$$
$$- a_{ij}(x_j - \ell_j) - a_{ik}(x_k = \ell_k) \leq M_i - b_i, \text{ if } a_{ij} < 0 \text{ and } a_{ik} < 0.$$

is valid and is not implied by $\ell_j \leq x_j \leq u_j$ and $\ell_k \leq x_k \leq u_k$.

For any two variables, we can collect all of the implied inequalities on those two variables and keep any that are not implied by the others. These inequalities can be combined to give transitively implied inequalities by adding multiples of two inequalities

$$\alpha_1 x_j + \alpha_2 x_i \leq \beta_1, \text{ and } \alpha_3 x_i + \alpha_4 x_k \leq \beta_2,$$

where $\alpha_2 > 0$ and $\alpha_3 < 0$. Cross multiplying to eliminate x_i gives

$$\alpha_1 x_j - \alpha_4 \frac{\alpha_2}{\alpha_3} x_k < \beta_1 - \frac{\alpha_2}{\alpha_3} \beta_2.$$

This new inequality may or may not be stronger than those already present for the pair of variables x_j, x_k. When it is stronger, it may give a tighter bound on one of the two variables. It may also be that $j = k$ so that the implied inequality is just a bound, which may or may not be tighter than already given. Any inequality derived or derivable in this way is called a transitively implied inequality.

Corresponding to the system of inequalities on two variables, we form a graph G with a node j for each variable x_j and an edge for each inequality. With the edge [j, k] corresponding to the inequality

$$\alpha_1 x_j + \alpha_2 x_j \le \beta$$

we associate three numbers: α_1 is the way that the edge meets node j, α_2 is the way that the edge meets node k, and β is associated with the edge itself. If the edge has one plus α_i and one minus α_i, then it is called directed, and its direction is from the end with a plus α toward the end with a minus α. We draw the edge with an arrow meeting node j if $\alpha_1 < 0$ and no arrow meeting node k if $\alpha_2 > 0$.

For 0-1 variables, the inequalities [13] are all of the form

$$x_j + x_k \le 1, x_j - x_k \le 0, \quad \text{or} \quad -x_j - x_i \le -1,$$

having $+1$ or -1 coefficients and 0, $+1$, or -1 right-hand sides. Then the edges of the corresponding graph meet nodes with either a $+1$ or a -1 value, and the right-hand side is not necessary to specify because the two coefficients determine it. Forming new inequalities is equivalent to forming the transitive closure [13] of this graph.

Returning to our case, the edge corresponding to a transitively implied inequality is called a transitively implied edge. The bounds on variables are considered to be one-ended edges at the corresponding nodes, lower bounds having a -1 coefficient at the node and upper bounds having a $+1$. These bounds can also be transitively implied, and such implied bounds are useful in identifying free variables, as is discussed below. See [11] for a development of the properties of this graph and its uses.

The case where variables x_i and x_j in an inequality are 0-1 variables is completely covered in [13]. A completely general analysis is given in [11], but a special case will be discussed here. Whenever setting a variable, continuous or integer, to one of its bounds implies that another variable must be at one of its bounds, an "implication graph" can be formed having directed edges with signs:

$$(+ i, +j) \text{ whenever } x_i = u_i \text{ implies } x_j = \ell_j ;$$
$$(+ i, -j) \text{ whenever } x_i = u_i \text{ implies } x_j = u_j ;$$
$$(- i, +j) \text{ whenever } x_i = \ell_j \text{ implies } x_j = \ell_j ;$$
$$(- i, -j) \text{ whenever } x_i = \ell_j \text{ implies } x_j = u_j ;$$

and nodes j for each variable x_j. Thus, edges have both a direction, showing the sense of the implication, and signs at the nodes showing upper or lower bounds. The direction of implication can be both ways, e.g. the inequality

$$\frac{x_i - \ell_i}{u_i - \ell_i} + \frac{x_j - \ell_j}{u_j - \ell_j} \leq 1 \text{ for lower and upper bounds } \ell_i < u_i, \ell_j < u_j,$$

gives both $(+ i, + j)$ and $(+ j, + i)$. The direction of implication is indicated by the ordered pair: $(+ i, + j)$ means $x_i = u_i$ implies $x_j = \ell_j$, and $(+ j, + i)$ means $x_j = u_j$ implies $x_i = \ell_i$. For 0-1 variables x_i and x_j, the implications are always both ways [13] because there are effectively only four inequalities over a pair of 0-1 variables x_i, x_j :

$$x_i + x_j \leq 1 , \ x_i - x_j \leq 0, - x_i + x_j \leq 0, \text{ and } - x_i - x_j \leq - 1.$$

When only one of x_i, x_j, say x_i, is 0-1, the implication will always be both ways and the form of inequalities possible are:

$$x_i + \frac{x_j - \ell_j}{u_j - \ell_j} \leq 1, x_i - \frac{x_j - \ell_j}{u_j - \ell_j} \leq 0, - x_i + \frac{x_j - \ell_j}{u_j - \ell_j} \leq 0, - x_i - \frac{x_j - \ell_j}{u_j - \ell_j} \leq - 1.$$

In continuous variables $0 \leq x_i \leq 1$ and $0 \leq x_j \leq 1$, for example, the inequality $x_i + 2x_j \leq 2$ leads to the edge $(+ j, + i)$ but not $(+ i, + j)$.

The signs on the nodes meeting an edge indicate the signs of coefficients α_i, α_j in an inequality of the form $\alpha_i x_i + \alpha_j x_j \leq \beta$. This convention is consistent with that for 0-1 variables [13].

In [2] it is shown that inequalities derived from edges of this graph can be effective in improving the linear programming bound especially for fixed change problems. However, there are other ways that building up and saving the implication graph could be useful. For example, coefficient reduction to be discussed in the next section can be improved. That and other constraints are discussed in the next section. Another way that implication graphs can be used is in the bound reduction and resultant fixing or freeing of variables. The idea is to strengthen Theorem 2.3 to say, e.g. where $a_{ik} > 0$ there, that if $\lambda_i = -\infty$ and $\overline{m}_i > b_i$, where $\overline{m}_i$ is the lower bound on $\sum a_{ij} x_j$ obtained as a result of setting $x_k = u_k$ and the resultant fixing of variables implied by the implication. Thus, the probing [10] for 0-1 variable can be extended to mixed 0-1 problems to build up the implication graph, but then the implication graph makes later probing easier to do.

Primal variables x_j might be fixed from optimality considerations, e.g. when a dual restriction $\pi A^j \leq c_j$ is strictly satisfied so is redundant then x_j can be fixed to its lower bound ℓ_j . This consideration applies to continuous as well as integer variables and is, in fact, a dual statement to Theorem 2.2. That is, dual testing and bound tightening can be done in a completely symmetric manner to fix primal variables.

Primal variables x_j may be made <u>free</u> whenever the lower and upper bounds, if initially given, are implied by the other constraints of the problem. This happens strictly if $\ell_j' > \ell_j$ or $u_j' < u$. The main advantage of identifying free variables is that they can be brought into the basis and left in the basis, knowing that even if they should go negative, they need not be dropped because they will eventually become non-negative again.

Another advantage to identifying a free variable x_j is that the dual constraint can be considered to be an equality $\pi A^j = c_j$ so that the bound tightening in section 2, applied to the dual problem, can use both $\pi A^i \leq c_j$ and $\pi A^j \geq c_j$ in order to reduce bounds on the dual variables. That is, we would like to use $\ell_j = -\infty, u_j = +\infty$ in the dual testing where it is advantageous to do so and at the same time use ℓ_j' , u_j' as implied by the other constraints in the primal bound tightening. In order to do so we must be sure that relaxing transitively implied bounds does not get into a cycle so that we are using a bound that has been relaxed in order to relax other bounds. This and other aspects of preprocessing are discussed in [11].

It should be evident that all of the above considerations apply equally to the dual problem. The proposal is that the problem be stated in a symmetric way so that the same code can be used for both the primal and dual problems.

Here, we mention some of the common features of a linear programming problem that a pre-processor looks for. A redundant constraint in the primal means that the constraint can either be deleted from the problem or it can be left in but the appropriate slack variable made basic and free (i.e. never dropped from the basis). However, a redundant constraint in the dual allows the corresponding primal variable to be fixed at one of its bounds.

Fixing a primal variable due to the implied bounds being so tight that the variable can only take on one value is obviously advantageous in that the variable drops from the problem. Fixing a dual variable π_i gets rid of a constraint. In practice one can either subtract a multiple of the row from the objective or make a slack variable t_i free with cost equal to the value at which the variable π_i is fixed.

Identifying a free variable in the primal allows us to pivot that variable into the basis and never drop it from the basis. Identifying free dual variable is useful also in that it allows the slack variables for that constraint to be dropped from the problem. Another advantage is that knowing that a primal constraint is an equation gives another inequality that one might use for tightening the primal bounds. However, in identifying free variables it is important to realize, as mentioned at the end of the preceding section, that one cannot just make free every variable whose bounds are transitively implied.

Integer variables x_j can be fixed whenever there is only one integer in the interval (ℓ_j, u_j) , e.g. a 0-1 variable can be fixed to zero whenever an upper bound smaller than one can be established.

Generally, the lower bounds on integer variables can be rounded up to the next closest integer, and upper bounds can be rounded down.

4. Constraint Generation

Coefficient reduction as given in [12] can be equally applied to mixed 0-1 problems. It is included here under constraint generation because the purpose is to tighten the linear constraints and improve the linear programming approximation to the mixed-integer programming problem. In essence, a uniformly stronger constraint is generated and replaces the original. The procedure, in general, can be applied whenever fixing a 0-1 variable to either zero or one would cause a constraint to become inactive. Suppose in the following that x_j is a 0-1 variable with non-zero coefficient a_{ij} in row i. The procedure is as follows:

$$\text{if } \lambda_i = -\infty \text{ and } a_{ij} < 0 \text{ and } M_i < b_i - a_{ij}, \text{ then change}$$

$$a_{ij} \text{ to } a'_{ij} = b_i - M_i;$$

$$\text{if } \lambda_i = -\infty \text{ and } a_{ij} > 0 \text{ and } M_i < b_i + a_{ij}, \text{ then change}$$

$$a_{ij} \text{ to } a'_{ij} = M_i - b_i \text{ and change } b_i \text{ to } b'_i = M_i - a_{ij};$$

$$\text{if } \mu_i = +\infty \text{ and } a_{ij} > 0 \text{ and } m_i > b_i - a_{ij}, \text{ then change}$$

$$a_{ij} \text{ to } a'_{ij} = b_i - m_i;$$

$$\text{if } \mu_i = +\infty \text{ and } a_{ij} < 0 \text{ and } m_i > b_i + a_{ij}, \text{ then change}$$

$$a_{ij} \text{ to } a'_{ij} = m_i - b_i \text{ and change } b_i \text{ to } b'_i = m_i - a_{ij};$$

The justification for these changes is the same as given in [12] and is that the modified constraints have the same solution sets in variables satisfying their lower and upper bounds and having x_j at either zero or one. The motivation for the change is always to tighten up the linear constraints. Coefficient reduction has been shown [17] to be of critical importance in fixed-charge problems. In [17], a more extensive procedure is used, and, in fact, a linear program is solved to reduce each of several coefficients. Even this much work seems to be justified for that class of problem.

The extension to special ordered sets also applies here since only lower and upper bounds on the variables outside of the special ordered sets was used (see section 4 of [12]). More generally, we can use the implication graph to improve the maximum M_i or minimum m_i. That is, if setting a 0-1 variable to zero or one allows the bounds on other variables to be tightened, or other 0-1 variables to be fixed, then the maximum M_i or minimum m_i on $\Sigma a_{ij}x_j$ might be smaller or larger, respectively, as a result, allowing more possibilities for coefficient reduction. In fact, zero coefficients might change to non-zeroes in this way. Such changes may be useful if the improvement in the linear programming approximation to the integer problem is sufficient to offset introducing more non-zero coefficients.

Several different types of constraints have been proposed and, in some cases, tested (see, e.g., [3] for application to the traveling salesman problem). One general class of stronger constraints is the 'disaggregated' [10] form of fixed charge models. The most common instance of a constraint that can be disaggregated is

$$\sum_{j \in T} x \leq (\sum_{j \in T} u_j)y, \text{ where } y \text{ is a } 0 - 1 \text{ variable,}$$

which can be replaced by the system of inequalities:

$$x_j \leq u_j y, \text{ for all } j \in T.$$

These latter, stronger, inequalities can be discovered from the implication graph and can be adjoined to the linear program when violated. Thus, one can be sure of only adding stronger inequalities in fact violated by some current linear programming solution. This approach is much in the spirit of cutting planes but the cuts here are sparse and can be very effective [15].

Another class of cuts successfully used [4] are the 'covering inequalities'. These can be also generated for mixed 0-1 problems by considering that an inequality $\Sigma a_j x_j \leq b$, for example, in 0-1 variables x_j for $j \in J_l$ and $\ell_j \leq x_j \leq u_j, j \in \{1, \dots, n\}\backslash J_l$ implies the inequality

$$\sum_{j \in J_l} a_{ij}x_j \leq b_i - \sum_{\substack{j \in J_i \\ a_{ij} > 0}} a_{ij}\ell_j - \sum_{\substack{j \in J_l \\ a_{ij} < 0}} a_{ij}u_j$$

For example, the inequality

$$x_l + 5x_2 + 6x_3 + x_4 + 3x_5 \leq 9,$$

with x_1, x_2, and x_3 required to be 0-1 and the other x_j non-negative, implies

$$x_1 + 5x_2 + 6x_3 \leq 9$$

from which can be implied $x_2 + x_3 \leq 1$ because both x_2 and x_3 equal to one is infeasible. In general, for an inequality in 0-1 variables $x_1, ..., x_n$

$$\sum_{j=0}^{n} a_j x_j \leq b \qquad \text{where } a_j > 0,$$

a <u>cover</u> is a subset J_c of the indices 1,...,n such that

$$\sum_{j \in J_c} a_j > b.$$

Then the cover inequality is simply

$$\sum_{j \in J_c} x_j \leq |J_c| - 1.$$

Finding a violated cover inequality is a knapsack problem [4]. In practice, any method can be used to solve that problem, and then the inequality can be 'lifted' [4] to a stronger inequality to try to identify a violated constraint valid for all 0-1 solutions.

In [2] it is shown that inequalities from implications together with tighter bounds can improve the linear programming relaxation and solution times for mixed 0-1 problems particularly of a fixed charge type. For example, fixing a 0-1 variable y_i to 0 may force several continuous variables to be 0, and then inequalities of the form $x_j \leq u_j y_i$ can be adjoined to the problem. The bound tightening procedure to reduce u_j can be useful here in order to adjoin tighter constraints. An example will be given to illustrate even stronger constraints that might be generated from the implication table. Consider the constraints:

$$u y_i \geq x_j, \ i = 1, 2 ; \ j = 1, 2, 3, \text{ where } u > 0,$$
$$-y_1 - y_2 \ \leq -1$$
$$y_1 \text{ and } y_2 \text{ 0-1}, \text{ and } 0 \leq x_j \leq u, j = 1, 2, 3.$$

In this mixed integer case, the implication graph includes the directed edges:

$$(-y_1, -y_2), \ (-y_2, -y_1), \ (x_j, -y_i), \text{ and } (-y_i, x_j), \ j = 1, 2, 3 \,; \ i = 1, 2.$$

Valid inequalities

$$\frac{1}{u} x_j - y_1 - y_2 \leq -1 \,, \ j = 1, 2, 3,$$

can be generated here even though x_j is not an integer variable. This constraint is a scaled version of the clique inequalities from [13]. The subgraph on the three nodes x_j, y_1, y_2 has all three edges present with the implications going both ways and the signs on the edges at any one node are all the same, for any node. More generally, if a constraint of the form

$$x_1 + x_2 \leq u,$$

is imposed, then a stronger clique inequality is implied;

$$\frac{1}{u} x_1 + \frac{1}{u} x_2 - y_1 - y_2 \leq -1,$$

but this observation does not extend to three variables, i.e. even if all three inequalities:

$$x_1 + x_2 \leq u_1, \ x_1 + x_3 \leq u, \text{ and } x_2 + x_3 \leq u$$

are present, one cannot derive the clique inequality:

$$\frac{1}{u} x_1 + \frac{1}{u} x_2 + \frac{1}{u} x_3 - y_1 - y_2 \leq -1$$

because $x_1 = x_2 = x_3 = u/2$ and $y_1 = y_2 = 1$ is valid and does not satisfy the above clique inequality. However, if $x_1 + x_2 + x_3 \leq u$ is one of the given inequalities, then the above clique inequality can be imposed.

Thus, the clique inequalities from [13] can be extended to the mixed 0-1 case provided the clique contains all implications in both directions, all signs the same at any given node, and no more than two continuous variables. For an extensive treatment of constraint generation in the mixed 0-1 case see [16].

5. Column Generation for Stronger LP

Some MIP problems are modeled in a way where the integer variables represent insignificant decisions in a larger decision problem. There are two problems that classically have been formulated in a better way and are always presented in that way. In this and the next two sections, an attempt is made to give a general framework for improving the linear programming relaxation for mixed 0-1 models by using column generation. The two cited examples will first be developed in a more naive formulation and then in the better, column generation formulation.

The first example we consider is the cutting stock problem [9]. Going back to its original form, the problem can be posed as follows: there are r rolls of paper of length $B_1, \dots, B_r$, and lengths $L_1, \dots, L_m$ to be cut from the rolls with demand $D_1, \dots, D_m$ for the m lengths. The version of the problem taken here for illustrative purposes is to have a fixed number of rolls of given lengths and to cut them so as to minimize the shortages in demand. An integer programming formulation is to let x_{ij} be the number of lengths L_i cut from roll j, and require:

$$x_{ij} \geq 0 \ \text{ and integer}$$

$$\sum_{i=1}^{m} L_i x_{ij} \leq B_j, \quad j = 1, \dots, r$$

$$\sum_{j=1}^{r} x_{ij} + s_i = D_i, \quad i = 1, \dots, m$$

$$\text{minimize} \sum_{i=1}^{m} s_i$$

Before proceeding, we should immediately caution that this formulation is not the way to address the cutting stock problem. Branch and bound would run forever on any reasonably large version of this problem. One way to state the reason is that the variables do not represent important decisions. For example, deciding that a given roll will not have any of a particular length cut from it has very little effect on the linear programming solution because that length can always be cut from some other roll. In a way, there is just too much symmetry in the formulation.

A better formulation is to let the variables be the pattern for cutting the rolls. That is, let y_j^k be a variable with corresponding column vector $a_j^k = (a_{1j}^k, \ldots, a_{mj}^l)^t$ where k indexes all solutions to

$$a_{ij} \geq 0 \quad \text{and integer,}$$

$$\sum_{i=1}^{m} L_i \, a_{ij} \leq B_j,$$

so that a_{ij}^k represents the number of lengths L_i cut from roll j in the k th solution to the problem. The y_j^k are, then, required to satisfy:

(5.1) $\quad y_i^k \geq 0 \quad \text{and integer,}$

(5.2) $\quad \sum_k y_j^k = 1, \quad j = 1, \ldots, r$

(5.3) $\quad \sum_{j, k} a_{ij}^k \, y_i^k + s_i = D_i, \, i = 1, \ldots, m$

while minimizing $\sum s_i$. There is one main reason for saying that this is a better formulation than the first one: the linear programming solution is closer to an integer solution. A main disadvantage is that there are many, many columns. Gilmore and Gomory showed how to overcome that difficulty by column generation: given an optimum linear programming solution for some subset of the columns where (π, σ) is the dual solution, σ_j being the dual variable for constraint (5.2) and π_i for constraint (5.3), solve the knapsack problem

$$\text{maximize} \sum_{i=1}^{m} \pi_i u_i \text{ subject to}$$

$$u_i \geq 0 \quad \text{and integer}$$

$$\sum L_i u_i \leq B_j,$$

and compare the objective value to σ_j. If less than σ_j, then the current linear programming solution is optimum. Otherwise, adjoin the new column having a 1 in row j of (5.2) and entries

$$a_{ij}^{k} \;=\; u_i$$

in the rows (5.3) for this new column k. The reason that this second linear program is better will be discussed in a general setting in the next section. For now, an example will be given to illustrate the improvement.

Suppose there are two rolls of lengths $B_1 = 8$ and $B_2 = 32$, two lengths $L_1 = 3$ and $L_2 = 9$ with demand $D_1 = 1$ and $D_2 = 4$. It is easy to see that there is no way to cut four lengths of $L_2 = 9$ from the two rolls of lengths 8 and 32. However, the first linear programming formulation is:

$$x_{ij} \ge 0 \quad \text{and integer}$$
$$3\,x_{11} + 9\,x_{21} \;\le 8, \quad 3\,x_{12} + 9\,x_{22} \le 32,$$
$$x_{11} + x_{12} + s_1 = 1, \quad x_{21} + x_{22} + s_2 = 4,$$
$$\text{minimize } s_1 + s_2,$$

which has an optimum solution $x_{11} = 1, x_{21} = 5/9, x_{12} = 0, x_{22} = 3\,4/9$, with $s_1 = s_2 = 0$. The second formulation, including all possible maximal columns, is:

$$y_j^{k} \ge 0 \quad \text{and integer}$$
$$y_1^1 + y_1^2 = 1, \quad y_2^1 + y_2^2 + y_2^3 + y_2^4 = 1$$
$$y_1^1 + 2y_1^2 + 10y_2^1 + 7y_2^2 + 4y_2^3 + 1y_2^4 + s_1 = 1$$
$$1y_2^2 + 2y_2^3 + 3y_2^4 \qquad + s_2 = 4$$
$$\text{minimize } s_1 + s_2$$

which has optimum solution $y_1^1 = 1, y_1^2 = 0, y_2^1 = y_2^2 = y_2^3 = 0$, and $y_2^4 = 1$. This solution has $s_1 = 0$ and $s_2 = 1$. Thus the linear program has an integer optimum. Of course, it is not generally true even for the stronger linear program that it will have an integer optimum. What is generally true, however, is that the first linear program will have objective value equal to zero provided the total lengths of the rolls is sufficient to meet the total of the demanded lengths. Thus, that linear program is worthless at providing an initial bound on the integer answer. In fact, it can remain worthless even after quite a bit of branching, due in part to the symmetry previously mentioned.

There is one remark to make here in relating this problem to the one treated in [9] : if all of the rolls have the same length, then the problem can be simplified to

$$y^k \geq 0 \text{ and integer}$$

$$\sum_k y^k = r$$

$$\sum_k a_i^k y^k + s_i = D_i, \, i = 1, \ldots, m,$$

$$\text{minimize } \sum_{l=1}^{m} s_l$$

The index j can be dropped because all of the rows are identical. The usual form is even simpler: require that the demand be satisfied and ask for the minimum number of rolls to meet that demand. The problem then has no variables s_l and has as objective to minimize r. The column generation problem in this formulation is the same as that before. The y^k represent the number of rolls cut using the "pattern" k, i.e. the kth knapsack solution telling how many to cut of each length. In fact, in the cutting stock problem it is desirable to have only a few different patterns, i.e. only a few y^k with non-zero values.

Gilmore and Gomory [9] suggest rounding up a variable close to integer until the linear program returns an integer solution. Such a simple rounding procedure is probably reasonably effective. In the next section, we discuss branching for 0-1 versions of the general problem treated by such a column generation procedure.

A second example of a problem where column generation is accepted is the crew scheduling problem. The problem is to assign crews of airplanes to scheduled flights. Although in practice, there may be several types of aircraft and several types of crews depending on the crew base and the qualification of the crew. There also may be complicated rules governing what makes up a legitimate tour for a crew. Let us give a simplified version of the problem to illustrate the point here. Suppose there are k crews, all of equal qualification and same base, to assign to flights $F_1, \ldots, F_m$ such that every flight gets a crew and every crew is assigned flights making up a feasible tour, where feasible tour must be defined. Assume that every flight has an origin airport with departure time and a destination airport with arrival time. In our simplified version of the problem, assume that a feasible tour is a collection of no more that four flights such that: (1) the flights can be ordered so that one flight arrives at the airport of the next's departure with at least one hour between; (2) the starting and ending airports are the same; (3) any flight in the tour can be designated as a deadhead flight for that crew; (4) the total time between start and finish cannot exceed 9 hours; (5) the total of the flight times for the non-deadhead flights cannot exceed 5 hours; and (6) the cost of a tour is a weighted sum of the deadhead flights. It should be clear that in practice the specifications of a tour can be quite complicated.

It would be possible to formulate an integer programming model of this problem with x_{ij}^t being 1 if crew j is assigned to flight i and y_{ij}^t being 1 if crew j deadheads on flight i, where the t indicates that the flight is the t th in this crew's tour. Every flight must be crewed so

$$\sum_{j,t} y_{ij}^t = 1, \quad i = 1, \dots, m,$$

and the cost is clearly $\sum c_i y_{ij}^t$. Capturing the rules for being a feasible tour can be done in this simple case, e.g. (5) is

$$\sum_{i,t} T_i x_{ij}^t \leq 5, \, j = 1, \dots, n.$$

where T_i is the flight time for flight F_i.

Perhaps it is needless to say that this formulation is not recommended. Yet one sees all too many integer programming models that look like this: hundreds of little 0-1 variables and thousands of special constraints. One other consideration is that the linear programming bound would not be good, and branching could go on forever. The recommended way to solve these problems is to separate the generation of feasible tours from the covering of flights. The problem then becomes one of covering flights using feasible tours. There are various codes for solving these problems, and some use is made of linear and integer programming. With improved mathematical programming codes, the use of them may increase.

The linear programming column generation approach consists of solving a linear program most of whose constraints correspond to flights and whose variables correspond to crew tours. The constraints are that the sum of the crews assigned to a flight must add up to exactly 1. There is also a constraint that the crews assigned cannot exceed the number k of crews. The objective is to minimize total dead-head cost. The linear program is solved for a given collection of tours, and then the dual variables on flights are used to evaluate possible new tours to enter. For this simplified model, the sum of the dual variables on flights covered in a tour must exceed the cost of dead-head portions of the tour plus the absolute value of the dual variable associated with the crew availability constraint. If so, then the tour is one that would enter into the linear programming solution basis. Thus, the column generation consists of looking for a feasible tour using the dual variables as returns on the flights.

6. A General Decomposition for MIP

The general problem considered is a mixed integer problem of the form:

$$x \geq 0 \text{ and integer, } \quad y \geq 0$$
$$Ax = b$$
$$Fx + Gy = g$$
$$cx + dy = z \text{ (min)}$$

Assume that the pure integer problem

$$x \geq 0 \text{ and integer}$$
$$Ax = b$$
$$cx = z \text{ (min)}$$

is relatively easy to solve, for any objective function c . In order for the linear programming approximation to be stronger, we need that the linear program is not too easy, that is, that the linear program does not give integer answers automatically. Otherwise, the objective function of the linear program will not be improved by the column generation approach to be given.

Assume that the pure integer problem has a finite number of integer solutions, say $x^1, ..., x^K$. Then the master problem is:

(6.1) $\lambda^k \geq 0$ and integer, $y \geq 0$,

(6.2) $\displaystyle\sum_{k=1}^{K} \lambda^k = 1$

(6.3) $\displaystyle\sum_{k=1}^{K} \left(Fx^k\right)\lambda^k + Gy = g$

(6.4) $\displaystyle\sum_{k=1}^{K} \left(cx^k\right)\lambda^k + dy = z \,(\text{min})$

Rather than list all integer solutions, which is usually a practical impossibility, they can be generated as dictated by the linear program. That is, given a subset of the integer solutions to the above linear program, use the dual variables π to (6.3) and solve:

$$x \geq 0 \text{ and integer,}$$
$$Ax = b$$
$$(c - \pi F)\, x = z\,(\text{min})$$

A solution x^* that has objective function value z^* satisfying

$$z^* < \sigma,$$

where σ is the dual variable corresponding to constraint (6.2), gives a column to enter into the basis of the master linear program.

We assume that the above pure integer problem is somewhat easy, e.g. the knapsack problem in the cutting stock problem. It does not matter how one solves the integer problem; in the cutting stock problem, for example, Gilmore and Gomory recommend a specialized dynamic programming approach to solve the knapsack problems that generate cutting patterns.

In case the pure integer problem has a block structure,

$$A = \begin{bmatrix} A^1 & & & & \\ & A^2 & & & \\ & & \cdot & & \\ & & & \cdot & \\ & & & & \cdot & \\ & & & & & A^L \end{bmatrix}, x = (x_1, \dots, x_L)$$

$$F = [F^1, \dots, F^\ell], c = (c^1, \dots, c^L)$$

each block can be solved separately, and then the master problem becomes:

$$\lambda_\ell^k \geq 0 \text{ and integer}, y \geq 0$$

$$\sum_{k=1}^{K} \lambda_\ell^k = 1, \ell = 1, \dots, L$$

$$\sum_{\ell=1}^{L} \sum_{k=1}^{K} (F^\ell x_\ell^k) \lambda_\ell^k + Gy = g$$

$$\sum_{l=1}^{L} \sum_{k=1}^{K} (c^\ell x_\ell^k) \lambda_\ell^k + dy = z \, (\text{min})$$

If two of the subproblem blocks are identical both in the pure integer subproblem and in how the solutions enter into the master problem, then they can be combined into one subproblem with the convex combination constraints

$$\sum_{k=1}^{K} \lambda_1^k = 1 \text{ and } \sum_{k=1}^{K} \lambda_2^k = 1$$

replaced by

$$\sum_{k=1}^{K} \lambda_3^k = 2.$$

The above case, however, leads to difficulty in the branching as is already the case for the cutting stock problem. We avoid making that substitution below in order to avoid difficulties in branching on the master linear program. So assume now that all of the convex combination constraints in the master problem are of the simple form

$$\sum_{k=1}^{K} \lambda_\ell^k = 1$$

and that there may be several of them. The subproblems may be pure integer or pure 0-1. That is, integer variables other than 0-1 variables may be present.

We now turn to the question of how to get integer answers to the master problem. Simply setting a variable λ_ℓ to 0 or 1 will not work since the subproblem would then have to try to avoid regenerating that solution, and algorithms for 2nd, 3rd, ... , kth best solutions would have to be faced, an obviously undesirable possibility. A better scheme is the following: if there are fractional values of λ_ℓ then there must be a component j of x_ℓ^k such that the convex combination

$$\sum_{k=1}^{K} x_\ell^k \lambda_\ell^k$$

has component j at a fractional value f. Branching can be accomplished by requiring x_ℓ to have upper bound $I_\ell(f)$, the integer below f, one one branch, and lower bound $I_u(f)$, the integer above f, on the other branch. This scheme then forces x_ℓ^k to have those bounds in the pure integer subproblem at that node of the branch and bound tree. Having a variable bounded is clearly preferable to having to check for prohibited solutions and continue as would be the case if a λ_ℓ is set to 0 on a branch. In any case, branching by bounding a component x_ℓ^k is better in that it more evenly partitions the solution set into two subsets.

Of course, it remains the case that in the pure integer subproblems, any solution method can be used. Also, the subproblems need not always be solved to optimality. If a solution is found that is good enough to enter into the master linear program , then the subproblem optimization can be stopped. The subproblem optimization may also be stopped in order to branch on the master problem, making the subproblems easier. This is, the master linear program can be branched on even though it is not optimized (because the pure integer subproblem may not have been optimized). No bound is then valid from the linear program, but bounds can be derived at the nodes created by branching, where the pure integer subproblem might be easier.

7. Application to Clustering

The problem we consider is to partition the nodes of an undirected graph G so that the node subsets satisfy some knapsack constraint and to minimize the sum of the costs on the edges going between subsets of nodes. Call the subsets of the partition clusters, and assume that the number of clusters is fixed, say equal to L. Assume also that the costs on the edges are positive. Let a_i be the weight associated with node i in the knapsack constraint and let b_ℓ be the capacity of cluster ℓ. Rather than minimize the costs between clusters, it is equivalent to maximize the sum of the cost on the edges inside the clusters. A straight-forward formulation of the problem is, then, to let $x_{i\ell}$ be 1 or 0 if node i is in or not in the ℓ th cluster. The integer programming model is:

$$(7.1) \quad x_{i\ell} \geq 0 \text{ and integer}$$

$$(7.2) \quad \sum_{i=1}^{n} a_i x_{i\ell} \leq b_\ell, \ \ell = 1, \dots, L$$

$$(7.3) \quad \sum_{\ell=1}^{L} x_{i\ell} = 1, \ i = 1, \dots, n$$

$$(7.4) \quad z_{ij}^\ell \leq x_{i\ell} \text{ and } z_{ij}^\ell \leq x_{j\ell}, \ i < j \text{ and } (i,j) \text{ an edge of } G$$

$$(7.5) \quad \text{maximize} \sum_{\ell=1}^{L} \sum_{i,j} c_{ij} z_{ij}^\ell$$

The similarity with the naive formulation of the cutting stock problem should be noted. The only differences with the cutting stock problem are: (1) the fact that here the $x_{i\ell}$ are required to be 0 or 1 where in the cutting stock problem they can be any non-negative integer; and (2) the way the objective value is determined. The z_{ij}^{ℓ} are variables introduced to model the objective function and are not required to be integer or non-negative because they will be automatically be so in any optimum solution provided the $x_{i\ell}$ are integer.

That this formulation has a weak linear programming relaxation can be shown for the special case where all $b_{\ell} = b$. In this case, note that in order for the problem to have a feasible integer solution it must be that

$$\sum_{i=1}^{n} a_i \leq Lb$$

Then, each $x_{i\ell} = 1/L$ and $z_{ij}^{\ell} = 1/L$ is a feasible solution to the linear program with objective value equal to $\sum c_{ij}$. The sum of the cost of edges between clusters is thus equal to zero. In a real sense, the linear program is worthless as a bound on the integer programming objective value.

If the problem is decomposed as in the preceding section with master problem give by (7.1) and (7.3), then the master problem is

(7.6) $\qquad \lambda_{\ell}^{k} \geq 0$ and integer,

(7.7) $\qquad \sum_{k=1}^{K} \lambda_{\ell}^{k} = 1, \ell = 1, \dots, L$

(7.8) $\qquad \sum_{\ell=1}^{K}\sum_{k=1}^{K} u_{i\ell}^{k} \lambda_{\ell}^{k} = 1, \; i = 1, \dots, n$

(7.9) $\qquad$ maximize $\sum_{\ell=1}^{L}\sum_{k=1}^{K} v_i^{k} \lambda \ell^{k}$

where each u_{ℓ}^{k} is a solution to the knapsack problem

$$u_{i\ell}^{k} = 0 \text{ or } 1,$$

$$\sum_{i=1}^{n} a_i u_{i\ell}^k \le b_\ell$$

and v_i^k is the maximum value of the objective junction

$$\sum c_{ij} z_{ij}$$

subject to

$$z_{ij} \le u_{i\ell}^k \text{ and } z_{ij} \le u_{j\ell}^k$$

The fact that this linear program might be a strong one is suggested by its similarity with the cutting stock problem. That it is strong can be shown explicitly for a special case that is admittedly rather special but for which there is no particular reason to suspect that it would be strong. Suppose that the graph is complete and has Lq nodes for L and q positive integers at least 2. Let all of the costs $c_{ij} = 1$. Then any partition of the nodes into L clusters is an optimum integer answer with objective value

$$L \ \frac{q(q-1)}{2}$$

The initial linear program still has objective value

$$\sum c_{ij} = \frac{L(L-1)}{2}$$

But the master linear program has the integer solution as its optimum as can be seen by the fact that

$$\sigma_\ell = \frac{q(q-1)}{2} \ , \pi_i = 0,$$

is a feasible and, in fact, optimum solution to the dual, where σ, π are dual variables to (7.7) and (7.8), respectively. It seems on the face of it that this problem is not so well suited to the decomposition because the knapsack problem is so trivial. However, no claim is made that the linear program will usually give integer answers.

In doing branch and bound on this master problem, there is a symmetry difficulty when the clusters all have equal capacity, and even more so when L is larger than 2 or 3. Using the symmetry, any one node can be put in the first cluster making it distinguished by that node. At any other point in the branching, if a node is specified to not be in the first cluster, then it can be put in the second cluster without loss of generality. In this way, one may be able to distinguish each cluster by some one node, which may be different in different parts of the branch and bound tree.

The subproblem is itself an interesting problem. For any ℓ , the subproblem is of the form:

$$u_i = 0 \text{ or } 1,$$

$$\sum_{i=1}^{n} a_i u_i \le b$$

$$z_{ij} \le u_i \text{ and } z_{ij} \le u_j$$

$$\text{maximize} \left(- \sum_{i=1}^{n} \pi_i u_i + \sum_{i,j} c_{ij} z_{ij} \right)$$

That is, there is a cost π_i for including a node, but there is a gain of c_{ij} whenever two nodes meeting an edge are included. Such a solution u_i gives a new column to enter the basis of the master problem if the objective value is larger than σ_ℓ, the dual variable to (7.7).

This problem is not an easy one. It itself can be addressed by again applying the decomposition method. Putting the knapsack problem in the subproblem and leaving the z_{ij} variables in this (sub-)master problem gives a master problem:

(7.10) $\lambda_k \ge 0$

(7.11) $z_{ij} \le u_i^k \lambda^k \text{ and } z_{ij} \le u_j^k \lambda^k$

(7.12) $\sum \lambda_k = 1$

(7.13)   maximize $\left(-\sum_{i,k} \pi_i \mu_i^k \lambda^k + \sum_{i,j} c_{ij} z_{ij} \right)$

and leads to subproblems of the form:

$$u_i = 0 \text{ or } 1$$

$$\sum_{i=1} a_i u_i \leq b,$$

$$\text{maximize} \sum_i (\tau_i + \mu_i) u_i,$$

where τ is the vector of dual variables for constraints (7.11).

In words, the master problem divides c_{ij} into two parts

$$\tau_i + \tau_j = c_{ij} \, , \quad i < j$$

and then gives a value

$$-\pi_i + \sum_{i < j} \tau_i + \sum_{h < i} \tau_n$$

to each node to use in solving the 0-1 knapsack problem. How strong the master problem is depends on how much of the difficulty is in satisfying the knapsack constraints and how much is in the nature of the costs. For example, where the knapsack constraint is just a cardinality constraint, i.e. has coefficients all equal to 1, the knapsack constraint is easy to satisfy, and perhaps the problem is better treated directly with, e.g., a suitably adapted constraint generation method.

References

[1]   Chamberlin, D.D, "SEQUEL 2: A Unified Approach to Data Definition, Manipulation, and Control", *IBM Journal of Research and Development* 20 [1976], 560-575.

[2]   Ciriani, T.A., S. Gliozzi, and E.L. Johnson, "Recent Advances in Pre-Processing Techniques for Large MIP Problems", to be presented at 13 MPS, Tokyo, August 1988.

[3]   Crowder, H., and M. W. Padberg, "Solving Large-Scale Symmetric Travelling Salesman Problems to Optimality", *Management Science* 26 [1980], 495-509.

[4]   Crowder, H., E. L. Johnson, and M. Padberg, "Solving Large-Scale Zero-One Linear Programs", *Operations Research* 31 [1983], 803-834.

[5]   Dolk, D.R., "Relational Data Models and Relational Data Base Systems", NATO ASI on Mathematical Models for Decision Support, Val d'Isere, France, July 26 - August 6, 1987. Proceedings to appear.

[6]   Esposito, M.C., and E.L. Johnson, "OASIS, A Modeling Environment in APL2 and SQL", to appear

[7]   Forrest, J., "Alternative Approaches to Modeling", TIMS/ORSA Joint National Meeting, Los Angeles, April 1986.

[8]   Geoffrion, A.M., "An Introduction to Structured Modeling", *Management Science* 33 [1987], 547-588.

[9]   Gilmore, P.C., and R.E. Gomory, "A Linear Programming Approach to the Cutting Stock Problem", *Operations Research* 9 [1961], 849-859.

[10] Guignard, M., and K. Spielberg, "Propagation, Penalty Improvement, and Use of Logical Inequalities", *Mathematics of Operations Research* 25 [1977], 157-171.

[11] Jensen, D. L., and E. L. Johnson, "Preprocessing and Implication Graphs for Mixed-Integer Programming Problems", [in preparation]

[12] Johnson, E. L., M. M. Kostreva, and U. Suhl, "Solving 0-1 Integer Programming Problems Arising from Large Scale Planning Models", *Operations Research* 33 [1985], 803-819

[13] Johnson, E. L., and M. W. Padberg, "Degree-Two Inequalities, C lique Facets, and Biperfect Graphs", *Annals of Discrete Mathematics* 16 [1982], 169-187.

[14] Lenard, M., "Representing Models as Data", *J. Management Information Systems* 2 [1986], 36-48.

[15] Mairs, T. G., G. W. Wakefield, E. L. Johnson, and K. Spielberg, "On a Production Allocation and Distribution Problem", *Management Science* 24 [1978], 1622-1630.

[16] Padberg, M. W., T. J. Van Roy, and L. A. Wolsey, "Valid Inequalities for Fixed Charge Problems", *Operations Research* 33 [1985], 842-861.

[17] Suhl, U. H., "Solving Large Scale Mixed-Integer Programs with Fixed Charge Variables", *Mathematical Programming* 32 [1985], 165-182.

[18] Tomlin, J.A., and J.S. Welch, "Formal Optimization of Some Reduced Linear Programming Problems", *Mathematical Programming* 27 [1983], 232-240.

Advances in Nonlinear Network Models and Algorithms

John M. Mulvey
Princeton University
School of Engineering and Applied Science
Department of Civil Engineering and Operations Research
Princeton, New Jersey 08544, USA

Abstract

Network models with nonlinear objectives occur in numerous economic, engineering and management applications. These areas are described, along with highly specialized nonlinear programming algorithms for solving the resulting large-scale problems. It is shown that nonlinear network models can be handled efficiently using serial or vector processing computers. Extensions to stochastic programs are discussed.

1. Introduction

The development of specialized algorithms for solving any particular class of optimization problems depends upon three primary issues. First, is the problem class wide enough to warrant the design of highly specialized tools when general purpose software is already available? Second, what is the speedup that results when specialized algorithms are executed? Third, do these codes give rise to the solution of problems that would not otherwise be solvable?

In the case of nonlinear networks, each of these questions leads to an affirmative answer with respect to a specialized implementation. The nonlinear network domain has a rich variety of applications ranging from air-traffic control to the estimation of social accounting matrices; the next section outlines several of these applications. Second, specialized software is more than an order of magnitude faster than general purpose coding and can handle very large problem instances. Based on these observations we may expect a high payoff from the effort involved in developing special software.

The nonlinear, generalized network optimization problem may be defined as follows:

[NLGN]

$$\text{Minimize} \quad f(\overline{x})$$

NATO ASI Series, Vol. F51
Algorithms and Model Formulations
in Mathematical Programming
Edited by Stein W. Wallace
© Springer-Verlag Berlin Heidelberg 1989

subject to

$$\sum_{j \in \delta_i^+} x_{ij} - \sum_{k \in \delta_i^-} m_{ki} x_{ki} = b_i, \quad \text{for } i \in N \tag{1}$$

$$l_{ij} \le x_{ij} \le u_{ij}, \quad \text{for } (i, j) \in E \tag{2}$$

where

$f(\bar{x})$ is a real-valued, continuously differentiable convex function

$\{N\}$ is the set of nodes in the network

$\{E\}$ is the set of arcs (edges) in the network

x_{ij} is the flow over arc (i, j)

m_{ki} is the multiplier on arc (k, i)

b_i is the supply or demand for node i

$\bar{x} = \{x_{ij} \mid (i, j) \in E\}$ is the vector of flows over all arcs in the network

$\delta_i^+ = \{j \mid (i, j) \in E\}$ is the set of arcs leaving node i

$\delta_i^- = \{j \mid (j, i) \in E\}$ is the set of arcs entering node i

$u_{ij}(l_{ij})$ are upper (lower) bounds on arc (i, j).

In matrix notation we have

[NLGN]

$$\text{Minimize } f(\bar{x})$$

subject to

$$A\bar{x} = b$$

$$l \le \bar{x} \le u$$

Any $\bar{x}$-vector satisfying constraints (1) and (2) is feasible; the entire feasible region is defined as X. The terminology *generalized* network results because of the presence of arc multipliers -- m_{ki}. This feature is important in modeling situations with gains or losses on the flow through an arc (e.g. in power generation planning) or whenever a change of units takes place in the flow between two nodes (e.g. in financial cashflow management). There are many instances of problems that can be represented as networks, some of which are summarized in Table 1 below.

Setting	Interpretation of			
	Nodes	Arcs	Potential Difference	Flow
Water/Gas distribution network	Pipe intersections	Pipes, pumps, valves	Potential difference	Flow of water-gas
Electrical networks	Connections	Wires, resistors, other devices	Voltage drop	Electrical current
Economics	Locations of supply or demand	Trade links	Price differences	Units of commodity
Communication networks	Computers, switches	Communications links	Time delay	Data
Road networks	Locations in in city	Roads	Time delay	Cars-trucks

Table 1: Some Network Optimization Models

An introduction to generalized networks and their use is given in Glover et al[19]. It is well known -- Dantzig[13] -- that the basis **B** of a generalized network can be written in the form

$$
B \ = \ \begin{bmatrix} B_1 & & & \\ & B_2 & & \\ & & \cdot & \\ & & & \cdot \\ & & & & B_p \end{bmatrix}
$$

where each B_i is either lower triangular or nearly lower triangular with only one off diagonal element. Using techniques from graph theory we associate rows of B_i with nodes and columns with arcs. Then every lower triangular submatrix B_i corresponds to a spanning tree with a root node and every submatrix B_i which is nearly triangular corresponds to a tree with a loop. Spanning trees with root nodes or with loops are defined as *1-trees*. Thus the basis is a *forest* of *1-trees*.

Generalized networks may be viewed as an extension of the class of *pure* networks where all multipliers are identically equal to +1. From a numerical point of view, however, the distinction between pure and generalized networks is an important one. The basis of a pure network is a spanning tree, not a forest, and this makes relevant computations very efficient. Also, since the constraint matrix is unimodular (i.e. all principal determinants are +1 or -1) one may use integer arithmetic in implementing pure network codes. It appears from current experience; see, for example, Brown and McBride[10], that generalized network codes are only marginally slower than their pure network counterparts on pure network problems. These codes, in turn, can achieve a hundredfold improvement in efficiency over general purpose linear programming software. For nonlinear network problems, however, this distinction ceases to be important since both applications require real arithmetic for their implementation. Naturally, the advantage of generalized network models is that they can represent a much richer class of applications.

The highly sparse nature of these problems allows for the solution of very large nonlinear networks. Several authors have designed nonlinear programming algorithms to exploit the special structure of pure networks. In a similar vein, the generalized network structure has been exploited in the context of specialized algorithms as discussed in Section 3.

The rest of this paper is organized as follows: Section 2 surveys different classes of [NLGN] applications, including references to the published literture. In Section 3 we discuss solution methodologies that are available for this problem class. Finally, Section 4 considers several generalizations of [NLGN] models to stochastic programs; conclusions and directions for future research are also included.

2. Nonlinear Network Applications

In this section we survey several prominent areas of application for the [NLGN] model. Most of the examples discussed here are large-scale and cannot be solved in a practical fashion with general-purpose, off-the-shelf software systems.

2.1. Scheduling and Planning

Two significant classes of [NLGN] applications from the scheduling domain are described in this subsection. Scheduling and planning models usually require several time-periods for their basic structure; it becomes a difficult task to define the correct length for the time-periods. If the periods are too short relative to the horizon, then the model becomes too large for current computers. If they are too long, then the lack of realism prevents the model from providing acceptable solutions. Recognizing that several scheduling and planning problems possess a network structure may have a significant impact on the solution of these problems. First, large-scale network applications can be solved easily and the level of detail provided in the model can be expanded considerably. Second, we may design planning tools with fast response time so that an interactive process (human scheduler-computer) becomes feasible.

2.1.1. Hydro-Electric Scheduling

Figure 1 depicts the reservoir system for the lower Tajo area in Spain. The system consists of nine dams and an auxiliary reservoir system. Every dam is represented by a set of nodes -- one for each time period in the model. Inflows to the reservoirs are obtained from long-term hydrological forecasting models. The necessary data for this model are the following:

- Network topology
- Upper and lower limits on the reservoir storage, turbine operations, pumping and spillage
- Initial and final levels of each reservoir
- Hydroelectric production coefficients for each reservoir as a function of its storage capacity and turbine operations.

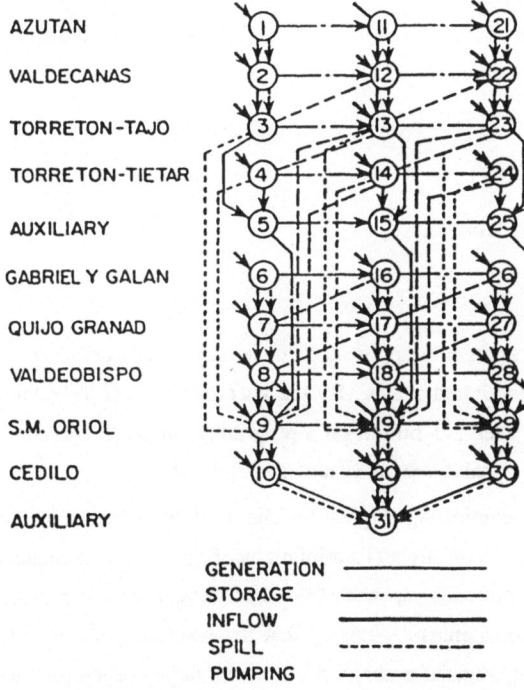

Figure 1: The Tajo Reservoir System

The conservation of flow equations are used to link the reservoirs to each other. For example, flow into J. M. Oriol is due to releases (via an auxiliary dummy reservoir) and spillage for power generation by Valdeobispo, Torreton-Tajo and Torreton-Tietar. J. M. Oriol releases its contents to Cedillo and it may be used to pump water to Tajo and Tietar. Generalized network multipliers can be used to account for losses of water in the different phases of the operation (e.g. evaporation during power generation or storage and absorption by the river bank during spillage).

This problem can become large rather quickly when used for long or even medium term planning. Typical applications also involve side constraints that may be nonlinear. In addition, multiple scenarios must be handled due to uncertainties in rainfall, temperature, demand, and so on. The largest problem solved so far, by Dembo et al.[14], was a nonlinear, nonconvex pure network problem with over 18,000 bounded variables and 5,000 network flow-conservation constraints.

The control of systems of dams in a river valley requires careful planning. There are numerous constituents who are affected by the results of decisions about dam releases. Not only does water impact irrigation, recreation and flood control, but also electrical power is generated through turbines driven by the released water. Rosenthal[38] has modeled the scheduling of reservoir releases by the Tennessee Valley Authority as a nonlinear network. Dembo et al[14]. used network-based planning tools for power system planning at Hidroelectrica Espanola. Other researchers have studied related aspects, including power-flow models[41, 42], optimal dispatch[6, 21], and hydrothermal dispatch[20]. Most of these problems can be represented as networks.

2.1.2. FAA Air-Traffic Control

One of the largest potential applications of nonlinear networks involves a planning-control system for air-traffic in the United States. By the year 2000, the U.S. Federal Aviation Administration (FAA) plans to spend over $11 billion on a program to upgrade and consolidate the U.S. air-traffic system -- called the National Airspace System Plan [NASP].

Monitoring and controlling flights on the high-level jet routes encompasses two (often) conflicting objectives. These are: (1) minimizing flight costs as measured by fuel consumption, enroute delays and airport delays, and (2) minimizing collision and other risks as measured by congestion. Mulvey and Zenios [31] showed that the resulting problem fits the [NLGN] framework. Unfortunately, the disaggregated problem is extremely large -- even for current supercomputers -- and further research is needed to develop practical optimization software.

Figure 2 depicts an example of a three-period operation model. Airports are represented by a set of nodes -- one for each time period. A plane may depart from Philadelphia (node PHL) for St. Louis (STL) at the scheduled time (T_0) or it may be delayed until T_1. Once it departs it may be assigned to alternative altitudes with varying characteristics with respect to fuel consumption and congestion. Once it arrives in St. Louis it may be granted permission to land right away or it may be delayed due to congestion.

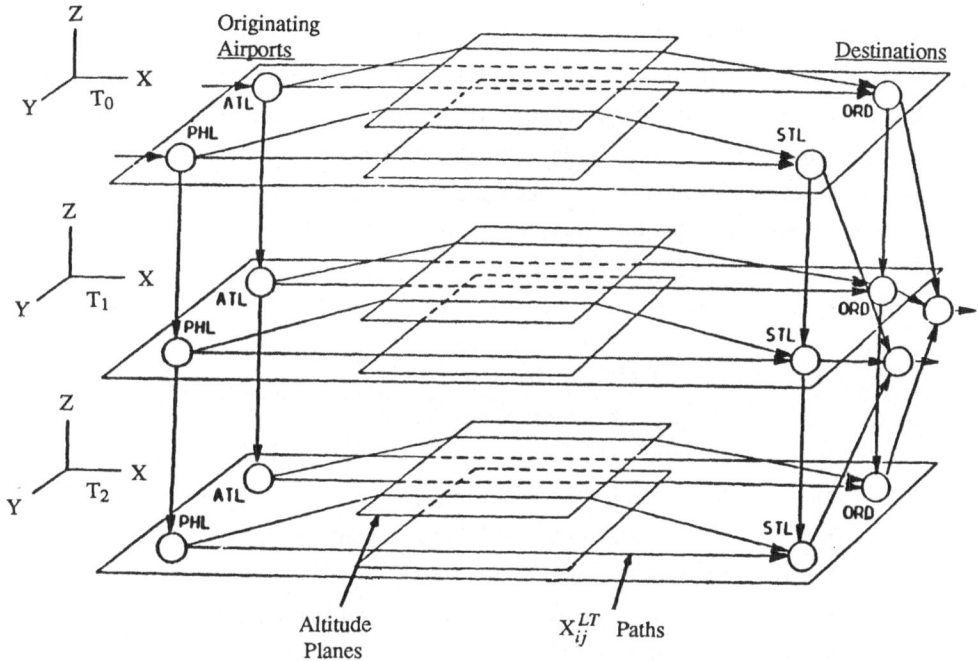

Figure 2: A Network Optimization Model for Air-traffic Control

2.2. Finance

Mathematical programming models have become an established part of financial analysis. Since the work of Markowitz[29] in the early 1950's, many of these models are based on nonlinear optimization. Several authors have proposed network models in which the arc multipliers play an important role. Rutenberg[40] and others suggested that multipliers can be used for translating currencies across countries (e.g. dollars to Deutschmarks). Crum and Nye[12], Barr[4], and Mulvey[33] designed multiperiod models in which interest, dividends and loans are modeled by means of multipliers. Cashflow management provides an ideal application for this methodology.

Nonlinearities arise in various contexts. The Markowitz model, for example, employs a quadratic objective function (portfolio variance) for measuring risks and returns. Constraints for this model guarantee conservation on the flow of funds. An important example is the decision involving the revision of a stock portfolio so that the expected value of the portfolio at the end of the planning horizon is maximized and so that a nonlinear risk function is minimized. Typically, risk is perceived as variation in return as measured by a quadratic function $\overline{y}^t Q \overline{y}$ where Q is the variance/covariance matrix and $\overline{y}$ is the vector of stock investments. It should be stressed, however, that the nonlinear network is not limited to the quadratic function. Higher order terms can be added, for instance, to represent the skewness of options portfolios and other financial instruments. In this single period model transaction costs are included as arc multipliers. Figure 3 depicts the network graph of this portfolio model. Note that the portfolio construction problem presents a special case.

The following notation and decision variables are required:

$\{I\}$	the set of stocks, with cardinality n
b_i	value of stock i in initial portfolio
b_0	cash available for investment purposes
t_{i1} (t_{i2})	arc multiplier for buy (sell) transaction of stock i
r_0	return multiplier for riskless asset
r_i	return multiplier for stock i held in portfolio
$F(\overline{y})$	variation of returns for portfolio $\overline{y}$
$l_i(u_i)$	lower (upper) bound on value of stock i in revised portfolio
s_i	value of stock i sold
p_i	value of stock i purchased
z_i	value of stock i maintained in portfolio
y_i	value of stock i in revised portfolio
x_n	amount invested in riskless asset
y_s	expected value of portfolio at the end of planning horizon.

The portfolio revision problem is now defined as:

Minimize $\quad \{w_2 \cdot F(\overline{y}) - w_1 \cdot y_s\}$

subject to

$$b_i - s_i - z_i = 0 \qquad \text{for all } i \in I$$

$$t_{i1} p_i + z_i - y_i = 0 \qquad \text{for all } i \in I$$

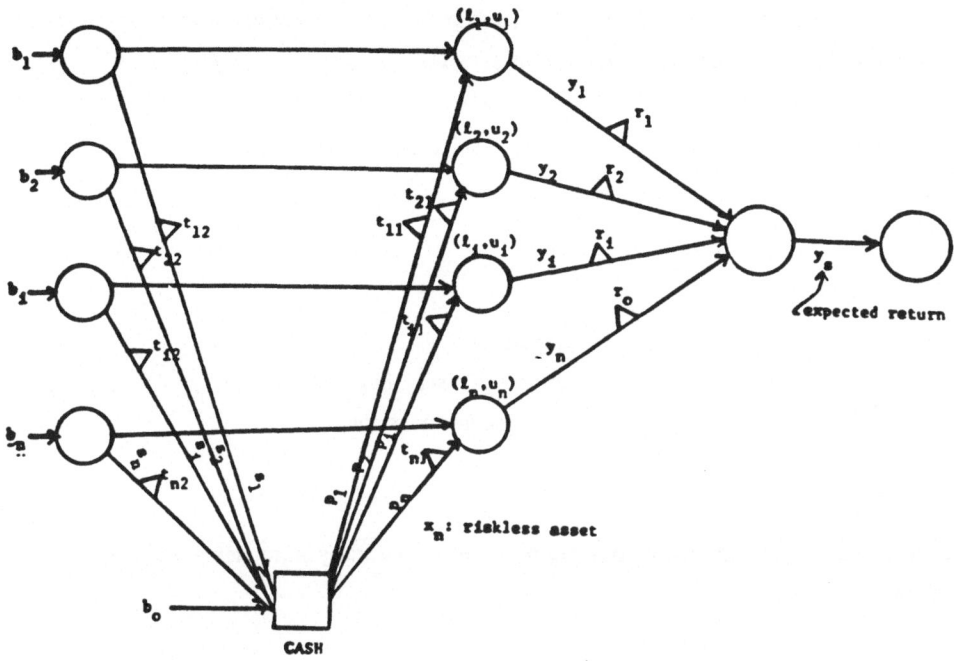

Figure 3: A Network Optimization Model for Portfolio Revision

$$\sum_{i \in I} r_i y_i + r_0 x_n - y_s = 0$$

$$b_0 + \sum_{i \in I} t_{i2} s_i - \sum_{i \in I} p_i = 0$$

$$l_i \le y_i \le u_i \qquad \text{for all } i \in I$$

$$w_1 + w_2 = 1$$

The goal programming weights (w_1, w_2) trace out the efficient frontier; the investor must select a compromise point on this frontier. This model is quite similar to the portfolio problem as presented in Rudd and Rosenberg[39] although they did not recognize the network structure.

Other practical aspects can be easily included in the nonlinear network by modifying the objective function. Often, the portfolio managers are provided with goals for the portfolio's volatility and yield. These goals are modeled by adding the following quadratic penalties to the objective function:

$$\lambda_c \cdot (\sum_j e_j y_j - E_T)^2$$

$$\lambda_p \cdot (\sum_j \beta_j y_j - \beta_T)^2$$

where

β_T	target beta
y_j	amount of asset j in portfolio
λ_c, λ_p	user specified constants
E_T	target yield

See the references for a discussion of other implementation issues surrounding the portfolio revision problem.

2.3 Equilibrium Models

There are many situations in which an equilibrium state on a network is defined by a solution to the following system of equations:

$$p_i - p_j \geq \phi_{ij}(x_{ij}) \qquad \text{if} \ \ x_{ij} = u_{ij}$$

$$p_i - p_j = \phi_{ij}(x_{ij}) \qquad \text{if} \ \ l_{ij} < x_{ij} < u_{ij} \qquad (3)$$

$$p_i - p_j \leq \phi_{ij}(x_{ij}) \qquad \text{if} \ \ x_{ij} = l_{ij}$$

and

$$\sum_{j \in \delta_i^+} x_{ij} - \sum_{j \in \delta_i^-} m_{ji} x_{ji} = b_i; \qquad i \in N \qquad (4)$$

where u_{ij}, l_{ij}, and b_i, $i \in N$ are given constants and $\phi_{ij}: R^n \to R$, $(i, j) \in E$. Inequalities (3) describe the relationship between potential drop across arc (i, j) and flow x_{ij}, whereas (4) describes flow conservation on the network. The value b_i may be interpreted as input (if positive) or withdrawal (if negative) at node i.

A network equilibrium problem is then defined as follows: Given the inputs/withdrawals b_i and limitations on flow, l_{ij} and u_{ij}, compute the potential drop and flows satisfying the equilibrium conditions (3) and (4). There are many instances of such models, some of which are summarized below:

- Traffic equilibrium, in which case arcs represent transportation paths (e.g. highways) and nodes are the connections between the paths.

- Market equilibrium, with nodes indicating spatially separated markets for products, and arcs indicating the interdependence between them.

- Demographic equilibrium, with nodes indicating distinct demographic zones and arcs indicating possible migration patterns between the zones.

- Water pipeline distribution systems, with arcs representing the pipes and nodes indicating the connection points.

- Electrical circuits, with arcs indicating devices or wiring and nodes indicating connections.

As described above, the potential drop across a specific arc (i, j) is affected only by the flow on this arc. There are more general models in which the potential drop could be a function of the flow on more than one arc (see the literature on variational inequalities for traffic equilibrium models). Such general models give rise to variational inequality problems which may or may not have nonlinear programming equivalents. However, the network equilibrium model described above is equivalent to a nonlinear optimization problem in the sense that its solution satisfies the Kuhn-Tucker necessary conditions of the following NLP.

$$\text{Minimize} \quad \sum_{(i,j) \in E} \int_0^{x_{ij}} \phi_{ij}(t)dt \tag{5}$$

subject to

$$\sum_{j \in \delta_i^+} x_{ij} - \sum_{j \in \delta_i^-} m_{ji} x_{ji} = b_i ; \quad i \in N \tag{6}$$

$$l_{ij} \leq x_{ij} \leq u_{ij} ; \quad (i,j) \in E \tag{7}$$

In most of the applications listed above the functions $\phi_{ij}(x_{ij})$ are monotonically increasing and hence the objective function (5) is convex and separable. Also, for the traffic and communication network applications, the flow conservation equations describe multi-commodity flow, where the potential drop depends on the total flow (sum of all the commodity flows on a link) (see, for example, Florian)[15]. Recently Hobbs[25] proposed some nonlinear network models for pricing in the electricity generation industry, but he solved these problems using linear approximations. Harker[22] proposed several network based models for studying spatial price equilibrium; he used nonlinear network technology to analyze available data. Also Barros and Weintraub[5] developed nonlinear network models to study the wheat market in Chile; the developed model is actually a generalized network with multipliers representing losses during overhauling.

Network equilibrium models are probably the largest nonlinear programming problems solved on a regular basis. In particular, the traffic equilibrium models, used for road and communication network planning, may have many hundreds of thousands of variables and constraints. The fact that they are multi-commodity problems and that there are usually no explicit upper bounds on the flows makes it possible to devise efficient, low-storage algorithms for their solution. In a sense, these are relatively easy problems. With the proliferation of computer networks and distributed processing, however, their importance and the importance of finding very efficient methods to solve them will grow.

2.4. Statistics and Large Databases

The timely collection of very large databases has become increasingly important over the past two decades. Many government agencies and private companies routinely depend upon these files for maintaining their operations. One important class of databases is known as microdata, whereby the file consists of a large number of individual decision units -- individuals, families, corporations, etc. Typically, micro-data files range in size from one thousand to over one million observations (decision units). Once collected, the raw data must be processed before it can be served up to the computer user. There are numerous steps involved, several of which employ network optimization. An important example is presented next.

2.4.1. Estimating Social Accounting Matrices

A very important tool used by economists in evaluating various sectors of an economy is the "input-output" matrix and its extensions of Social Accounting Matrices (SAM). An example of such a matrix is shown in Figure 4. The elements in the table correspond to the transactions between particular sectors. The essential characteristic of such matrices is that the row and the corresponding column sums must be equal.

	LAB	H1	H2	P1	P2	TOT
LAB		•	•	•	•	•
H1	•					•
H2	•					•
P1		•	•	•		•
P2		•	•	•		•
TOT	•	•	•	•	•	

(Blank entries indicate **possible** transactions)

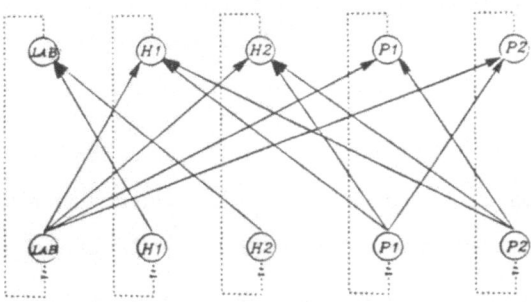

Figure 4: A Social Accounting Matrix and the Underlying Network Model

One problem encountered in using these tables is the following. The elements in the table are usually calculated from different agents in an economy and are often obtained through sampling procedures. The total expenditure (or income) of the agents, however, is normally available through government sources and is in general more accurate and up-to-date.

The model may thus be formulated as follows: given a SAM whose elements are out-of-date, compute updated values for these elements that satisfy a prespecified set of row and column sums. Such problems arise not only in economics but also in contingency table analysis, in statistical applications, in the analysis of congestion in telecommunication networks and in demographic studies in the social sciences. Mathematically, the problem may be formulated as shown below. Let x_{ij}^{old} be the given entries in the matrix, and let r_i denote the given new row sum for the ith row and similarly use c_j to denote the jth column sum. The new entries in the matrix must therefore satisfy

$$\sum_j x_{ij} = r_i \ (row \ sum)$$

(8)

$$\sum_i x_{ij} = c_j \quad (column \ sum) \tag{9}$$

$$x_{ij} \geq 0 \ for \ (i, j) \ in \ the \ matrix \tag{10}$$

The manner in which the new entries are postulated to relate to the old entries depends on the behavioral assumptions that are most plausible in the particular model under study. We will give an example:

Find x_{ij} as "close as possible" to x_{ij}^{old}. Using a least square criterion this may be formulated as

$$\underset{x}{Minimize} \ \sum_{ij} (x_{ij} - x_{ij}^{old})^2 \tag{11}$$

Thus (8) to (11) constitute a quadratic optimization model with network flow conservation (transportation) constraints. Introducing a weighting mechanism of the form $\frac{1}{x_{ij}^{old}} \cdot (x_{ij} - x_{ij}^{old})^2$ does not change the quadratic nature of the model. The optimal solution will be a chi-square estimate of the new matrix.

2.5 Others

There are numerous applications which are currently solved as large-scale networks. Many of these could benefit by adding nonlinear capabilities. Two examples are presented in this subsection. Others will undoubtedly arise as engineers become aware of the capabilities of nonlinear optimization software and the availability of high level modeling languages like GAMS of Bisschop and Meeraus[9].

2.5.1. Distribution and Logistics

Models of integrated production, inventory, and distribution are often based on a network form. The flow of materials from the factory through distribution centers and on to customers presents an appealing case where much of the problem can be depicted on a graph. The commensurate gain in efficiency allows the modeler room for including and modeling features, such as more time periods, disaggregate customer zones, additional potential sites for warehouses or a more realistic cost structure. Furthermore, penalty methods and decompositions can be easily incorporated into the [NLGN] solution strategy. These concepts would address side constraints.

2.5.2. Personnel Planning

Figure 5 presents an example of a multi-period network model for personnel planning. Here, people flow through the network from left to right -- for time -- and from top to bottom -- for employee rank. Note that new employees enter the model through the h variables, based on their background and (possibly) salary or experience levels. Attrition is modeled by means of the loss multipliers on arcs connecting the adjacent periods, e.g. l_{12} represents the loss ratio for job category 1 in time period 2. These values are generally based on historical data. Lower bounds on the employment arcs assume that a proper personnel force is available as needed. Upper bounds on hiring arcs indicate the scarcity of certain job skills in the work force or, perhaps bottlenecks in the hiring division. Refer to the paper by Mulvey[32] for additional details.

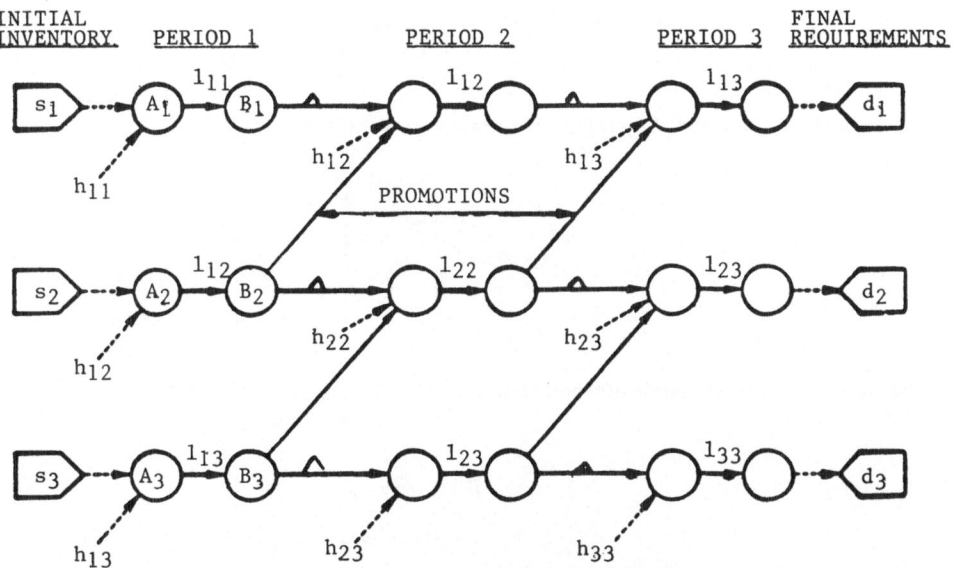

Figure 5: Multi-period Personnel Planning Model

3. Nonlinear Network Algorithms

This section takes up the topic of tailoring nonlinear programming algorithms for the network constraint structure. We have designed two separate nonlinear network algorithms: (1) truncated-Newton, and (2) simplicial decomposition. Next, we provide a brief overview of the truncated-Newton [TN] procedure.

The TN algorithm is a linesearch-based active set method. It consists of a sequence of iterates in each of which a feasible descent direction and a stepsize along this direction are computed. The algorithm requires an initial feasible starting point which may be obtained by linear programming or, alternatively, it may be started at an arbitrary point with artificial variables added. At each stage the stepsize is computed so as to ensure feasibility and thus a sequence of feasible points are generated. Before describing how the feasible directions are computed we need to introduce some notation.

Let A be partitioned as $[B \mid S \mid N]$ where B is a nonsingular basis matrix and let all vectors related to the variables $\overline{x} \in R^n$, such as $g(\overline{x})$, the gradient of f; and $p \in R^n$, a search direction, be partitioned in a corresponding manner, i.e.

$$\overline{x} = \begin{bmatrix} \overline{x}_b \\ \overline{x}_s \\ \overline{x}_n \end{bmatrix} \qquad g(\overline{x}) = \begin{bmatrix} g_b(\overline{x}) \\ g_s(\overline{x}) \\ g_n(\overline{x}) \end{bmatrix} \qquad p = \begin{bmatrix} p_b \\ p_s \\ p_n \end{bmatrix}$$

The Hessian of f is denoted by $H(\overline{x})$. We next define the projection matrix Z

$$Z = \begin{bmatrix} -B^{-1}S \\ I \\ 0 \end{bmatrix}$$

The truncated-Newton search direction is then

$$p = Zp_s \tag{12}$$

with p_s computed by solving the reduced-Newton equations

$$(Z'H(\overline{x})Z)p_s = -Z'g(\overline{x}) \tag{13}$$

inexactly using a linear conjugate gradient procedure.

While the benefits of the Newton direction p are greatest in the neighborhood of a solution, it is expensive to calculate the solution of equation (13). In response, we adjust in a dynamic fashion the degree of accuracy in solving the reduced Newton equations (13). A forcing sequence $\{\eta\} \to 0$ is employed. It reduces η from an initial value to a specified final value in a number of steps such that at each step the ratio of values of η are one half the previous ratio. Accuracy is defined according to the relative residual $||r||/||Z'g(\bar{x})||$ in which $r = (Z'H(\bar{x})Z)p_s + Z'g(\bar{x})$. Minor iterations continue only until the required accuracy is attained. Thus, $||r||/||Z'g(\bar{x})||$ defines the termination criteria for the minor iterations.

The algorithm has several important features:

(i)   it may be shown to be convergent under a nondegeneracy assumption;

(ii)   it will solve nonconvex, nonseparable problems (since the truncated-Newton direction obtained by solving the reduced Newton equation inexactly does not require the reduced Hessian to be positive definite);

(iii)   it does not require knowledge of the Hessian of $f(\bar{x})$ since the reduced-Newton equations may be solved by a preconditioned conjugate gradient algorithm which only requires products of the form $(Z'HZ)\bar{v}$ which may be obtained by differencing the gradient along the direction of $\bar{v}$;

(iv)   for the same reasons as in (iii), the matrix $Z'HZ$ (which is usually dense) need not be formed explicitly as it would have to be if a direct equation solver were used;

(v)   a conjugate gradient method "solves" the reduced Newton equations; thus, the storage requirements are of the order of a few vectors of dimension $||s||$.

The simplicial decomposition [SD] algorithm has also been specialized for generalized network constraints. SD has several origins, including Holloway[28], and Von Hollerbalken[26, 27] who named the procedure and established finite convergence properties.

The algorithm can be explained as a sequence of master problems [MP] followed by subproblems [SP]. The former takes a set of corner points of X, called a generating set $\{\bar{x}^1, \bar{x}^2, \ldots, \bar{x}^k\}$ and solves the resulting essentially unconstrained nonlinear program:

[MP]

$$\text{Minimize} \quad f(\lambda_1\bar{x}^1 + \lambda_2\bar{x}^2 + \cdots + \lambda_k\bar{x}^k)$$

subject to

$$\lambda_1 + \lambda_2 + \cdots + \lambda_k = 1$$

$$\lambda_i \geq 0 \qquad \text{for } i = 1,\ldots, k.$$

The [MP] can be optimized efficiently as long as the size of the generating set is relatively small ($k \leq 50$ or so).

Additional corner points are derived by solving the subproblem as a linear generalized network. A first order Taylor series $[f_L(\overline{x})]$ approximation at the current iterate replaces the nonlinear function $f(\overline{x})$. The subproblem becomes: minimize $f_L(\overline{x})$ subject to $\overline{x} \in X$. This linear network problem is easy to solve quite efficiently, using previous vertices as starting points and other numerical devices. See Ahlfeld, Mulvey and Zenios[30] for further details of the algorithm. Also see Hearn et al[23]. for a restricted [SD] version.

To illustrate the relative efficiency of specialized codes, we have compared the nonlinear network program NLPNETG[3] with the general purpose procedure MINOS on a common set of test problems (see Table 2). All of these problems were solved on a VAX 11/750 at Princeton University, using standard coding parameters. It would be noted that several of these test cases could not be solved in a practical fashion with MINOS. Table 3 lists some details of the test problems.

It is evident that the NLPNETG package is much more efficient than MINOS. Of course, we would expect some improvements simply because of the network structure. Still, the efficiencies are quite considerable, especially when put in the context of the wide domain of application for [NLGN].

Problem	NLPNETG Solution		MINOS Solution	
	Time (sec)	l_∞ norm	Time (sec)	l_∞ norm
PTN30	7.33	4.5E-3	44.08	8.8E-6
PTN150	23.86	2.0E-3	305.01	8.8E-6
PTN660	297.93	1.6E-2	9647.35	4.0E-4
SMBANK	21.50	3.0E-4	287.20	4.9E-4
BIGBANK	9100.00	4.0E-4	28800.0*	4.0E-1
RANBANK	245.37	4.0E-1	5477.97*	1.0E+2
STICK1	19.05	1.0E-4	12392.00	3.9E-6
STICK2	103.12	1.0E-4	28800.0**	
STICK3	73.75	6.0E-4	28800.0**	
STICK4	166.50	1.0E-3	28800.0**	
GROUP1ac	1652.00	6.0E-3	2312.69	5.9E-2
GROUP1ad	10227.00	4.9E-3	20347.00	5.9E-1
GROUP1ae	7376.00	4.1E-1	16115.00	2.8E-1
MARK1	3.07	2.0E-3	12.23	1.3E-1
MARK2	21.60	2.4E-1	33.03	1.4E-1
MARK3	204.43	7.1E-2	1341.53	1.5E-1
FAA.TEST	1.24	5.0E-2	15.22	3.4E-9
FAA.IND3	707.00	5.0E-2	111600.00	2.8E-4

* Program terminated prematurely; objective value within 0.02% from optimum
* * Program stopped by user after 8 hours of CPU time

Table 2: Comparison of NLPNETG with MINOS

PROBLEM	Size (Nodes/Arcs)	Free arcs at optimum	Condition No. of Reduced Hessian	Objective Function Value	Description
PTN30	30/ 46	15	$\geq 10^4$	-.3239322E5	Dallas Water
PTN150	150/ 196	44	$\geq 10^4$	-.4819730E5	Distribution
PTN660	666/906	240	$\geq 10^6$	-.2061074E6	models
SMBANK	64/ 117	54	$\geq 10^4$	-.7129290E7	Matrix
BIGBANK	1116/2230	946	$\geq 10^8$	-.4205693E7	balancing
RANBANK	800/2000	13	$\geq 10^4$	.3018788E6	models
STICK1	209/ 454	246	$\geq 10^2$	.6934392E1	Stick
STICK2	650/1412	763	$\geq 10^2$	.3124563E1	percolation
STICK3	782/1686	905	$\geq 10^2$	.1117978E2	models,
STICK4	832/2264	1433	$\geq 10^2$	.1566195E1	electrical networks
GROUP1ac	200/ 500	100	$\geq 10^3$	.1011792E5	Randomly
GROUP1ad	400/1000	119	$\geq 10^4$	.3834884E5	generated,
GROUP1ae	800/2000	230	$\geq 10^5$	.2265714E6	strictly convex networks
MARK1	23/ 42	4	$\geq 10^2$	-.1145214E6	Markowitz
MARK2	53/ 102	10	$\geq 10^2$	-.3271077E6	portfolio
MARK3	857/1710	35	$\geq 10^2$	.1150057E8	construction
FAA.TEST	28/ 64	6	$\geq 10^2$	.7725644E5	Air traffic
FAA.IND3	1295/2873	787	$\geq 10^2$	.2556325E1	control model

Table 3: Test Problems

4. Extensions

In this section we discuss an extension of the [NLGN] methodology to problems with stochastic coefficients. We are particularly interested in two period problems that have simple or network recourse.

When linear generalized network problems are coupled with independent stochastic demands and simple recourse, the resulting model becomes a nonlinear network with the following structure:

[ISN]

$$\text{Minimize} \ \ P(\overline{x}) \ = \ c'\overline{x} \ + \ \sum_{k \in T} \sum_{j \in N^k} \int_0^{y_j^k} c_o^k \cdot (y_j^k - d_j^k) \ \cdot \ dF_j^k(d_j^k)$$

$$+ \ \sum_{k \in T} \sum_{j \in N^k} \int_{y_j^k}^{Y_j^k} c_u^k \cdot (d_j^k - y_j^k) \cdot dF_j^k(d_j^k)$$

where

$$y_j^k = \sum_{i \in I_j^k} x_{ij}^k \quad \text{for } j \in N^k \text{ and } k \in T$$

$T \qquad\qquad$ set of periods in planning horizon

N^k set of demand nodes in period $k \in T$

I_j^k set of nodes connected to demand node $j \in N^k$, period $k \in T$

$F_j^k(\cdot)$ distribution function for demand at node $j \in N^k$, period $k \in T$

c_o^k per unit overage cost, period $k \in T$

c_u^k per unit underage cost, period $k \in T$

Y_j^k maximum value of demand at node $j \in N^k$, period $k \in T$

In this model, we have added a new set of decision variables -- y_j^k -- representing the total amount of flow provided to demand node j in time period k. Since the demands are uncertain, the model includes the expected costs of overage and underage along with the other costs in the objective function. A similar formulation was proposed in the context of transportation problems with uncertain demands by Cooper and Leblanc[11]. In their research the Frank-Wolfe algorithm was modified for solving the resulting nonlinear network model. Instead, we propose to adapt the truncated-Newton method[3] as discussed in Section 3. Also see Qi[35] and Ziemba[47] for specialized algorithms for this stochastic programming problem.

In the second stochastic network, we drop the independent random variable assumption. For many planning problems, it is unrealistic to assume that the inter-period forecasts are independent. For instance a lower than expected demand in period k is likely to lead to a lower than expected demand in period $k + 1$.

The objective function for the resulting nonlinear optimization model becomes:

[DSN]

Minimize $P(\overline{x}) = c'\overline{x}$

$$+ \sum_{k \in T} \sum_{j \in N^k} \int_0^{y_1^k} \int_0^{y_2^k} ....c_o^k \left[(y_1^k - d_1^k) + (y_2^k - d_2^k) + \cdots \right] dF^k(d_1^k, d_2^k, ...)$$

$$+ \sum_{k \in T} \sum_{j \in N^k} \int_{y_1^k}^{Y_1^k} \int_{y_2^k}^{Y_2^k} ....c_u^k \cdot \left[(d_1^k - y_1^k) + (d_2^k - y_2^k) + ... \right] dF^k(d_1^k, d_2^k, ...)$$

where $F^k(\cdot)$: conjoint distribution function for demands in period $k \in T$. The objective function is defined so that the dependencies among the random demands form a sparse pattern; otherwise, it would be difficult to specialize the optimization algorithm.

What are the alternative approaches for solving [ISN] and [DSN]? The most straightforward idea is to employ a general purpose nonlinear optimizer, such as MINOS[34] or GRG;[1] unfortunately, these methods are impractical for realistic size problems. Also in the case of [DSN], a functional approximation is likely to be inaccurate given the multiple integral structure of the objective function. In any event, solution of this model requires greater care, since even the evaluation of $P(\bar{x})$ at a single location can be time consuming. The algorithm should take advantage of any conditional independent relations -- by means of the partition of the $\bar{x}$ variables into $[\bar{x}_b \mid \bar{x}_s \mid \bar{x}_n]$.

Going beyond simple recourse, we note that several researchers have made contributions to the problem of stochastic programs with network recourse. See Birge[7], and Wallace[45,44].

In the models described above the uncertain quantities are restricted to the right-hand-side vector of the constraint equations. The resulting programs are solved by incorporating in the objective function penalty terms representing the total expected overage and underage cost. However, in a number of other important applications uncertainties could be encountered in the coefficients of the constraint matrix. In these cases, the uncertainty about problem parameters in each period is typically modeled by a relatively small number of joint realizations (scenarios) for the stochastic quantities. As an example, consider a two-period model for the portfolio investment problem in which the returns of risky assets, as well as cash inflows/outflows, are treated as stochastic quantities. The deterministic equivalent program to the underlying multiscenario problem becomes:

$$\text{Maximize} \quad \sum_{f \in F_t} \sum_{d \in D/_{t+1}} \pi_f \, \pi_d \, U(W/_{t+1}^d) \tag{14}$$

subject to

$$K_{it} + S_{it} = B_{it} \quad \text{for all } i \in I \tag{15a}$$

$$K_{bt} + S_{bt} = B_{bt} \tag{15b}$$

$$\sum_{i \in I} S_{it}(1 - C_{it}^s) - \sum_{i \in I} P_{it} + P_{bt} - S_{bt} - P_{1t} - P_{2t} + H_t = 0 \tag{16}$$

$$\left[K_{it} + P_{it}(1 - C_{it}^p) \right](1 + R_{it}^f) - K_{it+1}^f - S_{it+1}^f = 0 \quad \text{for all } i \in I, f \in F_t \tag{17a}$$

$$(K_{bt} + P_{bt})(1 + R_{bt}) - K_{bt+1}^f - S_{bt+1}^f = 0 \quad \text{for all } f \in F_t \tag{17b}$$

$$\sum_{i \in I} S_{it+1}^f (1 - C_{it+1}^s) + P_{1t}(1 + R_{1t}) - \sum_{i \in I} P_{it+1}^f + P_{bt+1}^f - S_{bt+1}^f$$

$$- P_{1t+1}^f - P_{2t+1}^f + H_{t+1}^f = 0 \quad \text{for all } i \in I, f \in F_t \tag{18}$$

$$W_{t+1}^{fd} = \sum_{i \in I} \left\{ \left[K_{it+1}^f + P_{it+1}^f (1 - C_{it+1}^p) \right] (1 + R_{it+1}^{fd}) \right\} + P_{2t}(1 + R_{2t})^2$$

$$+ P_{1t+1}^f (1 + R_{1t+1}^f) + P_{2t+1}(1 + R_{2t+1}^f) \tag{19}$$

$$- \left[K_{bt+1}^f + P_{bt+1}^f \right] (1 + R_{bt+1}^f) \quad \text{for all} \ f \in F_t, \ d \in D_{t+1}^f$$

where

$\{I\}$	set of risky assets
B_{it}	initial holdings (deficit) in risky asset $i \in I$
B_{bt}	initial liability (borrowing debt)
$C_{it}^s \ (C_{it}^p)$	per unit transaction cost for selling (purchasing) asset $i \in I$ at time t
R_{bt}	interest rate for borrowing in period t
$R_{1t}(R_{2t})$	rate of return for riskless asset maturing in one (two) period(s) when purchased at time t
$S_{it}(P_{it})$	amount of risky asset $i \in I$ sold (purchased) in period t
K_{it}	amount of risky asset $i \in I$ carried forward at time t
$S_{bt}(K_{bt})$	amount of liability paid back (carried forward) at time t
P_{bt}	amount borrowed in period t
F_t	set of scenarios for uncertain parameters in period t
D_{t+1}^f	set of scenarios for uncertain parameters in period $t+1$ when scenario in period t is $f \in F_t$
R_{it}^f	rate of return for risky asset $i \in I$ at time t for first period scenario $f \in F_t$
W_{t+1}^{fd}	net wealth at the ned of the second period ($t+1$) under first period scenario $f \in F_t$ and second period scenario $d \in D_{t+1}^f$
$U(.)$	appropriate utility function (typically concave).

The remaining variables and parameters are similarly defined.

The objective in the above program is the maximization of the utility of the net wealth at the end of the second period ($t+1$) under the postulated scenario structure. The constraints in (15) balance first period transactions to the initial portfolio composition while (16) ensures balance in the sources and uses of funds in period t. Equations (17) impose continuity in assets and liabilities between periods t and $t+1$ and (18) dictates balance in the sources and uses of funds in period $t+1$ for all first period scenarios $f \in F_t$. Finally, the equations in (19) express the resulting net wealth at the end of the second period in terms of the transaction policies and the returns of risky assets under the postulated scenarios.

In the two period model described above the constraint coefficient matrix has a dual block diagonal structure. In the case of linear programming, this structure can be exploited by means of decomposition[24, 18, 2, 46], basis factorization[16, 17, 8], or partitioning[37]. The work of Scott[43] provides evidence that extensions of these methods to nonlinear programs might be plausible. However, given the fact that the size of the problem increases exponentially with the number of scenarios, explicitly solving the deterministic equivalent program is not likely to be the best line of attack. It should be further noted that when a larger number of periods is considered, the constraint coefficient matrix has an arborescent structure (a special sparce block pattern of the block triangular structure) which cannot be exploited as easily.

A different approach proposed by Rockafellar and Wets[36] decomposes the program into a number of independent scenario subproblems, which in this case have the form of nonlinear generalized networks. This so called scenario aggregation procedure is particularly appropriate for problems that possess stochastic variables with a large degree of correlation, such as those occurring in financial planning models (see Section 2.2). The network structure is preserved during the algorithm; subproblems can be solved in parallel. Undoubtedly, research in this area will continue as computers become cheaper and more powerful and the need for solving stochastic networks grows as well.

5. Discussion

This section highlights some of the questions and answers that took place following Professor Mulvey's lecture at the NATO Conference.

Q. I am confused by the use of arc multipliers in your financial models. Why would the multipliers be needed if you held the S&P 500 stocks? What about the issue of holding cash?

A. In the case of the S&P 500, the multiplier reflects dividends that occur on stocks held within this group during the time period under study. In the case of cash, you would utilize a money market instrument and the multiplier would depict expected interest earned.

Q. What happens if everyone begins using network models for identifying arbitrage opportunities in financial markets?

A. If everyone used optimization, there would be few occasions when arbitrage took place. Markets would be in equilibrium for longer periods of time than what you see occurring today.

C.   My comments are on one of the applications that you mentioned, i.e., the balancing of social/national accounting matrices. This problem, unlike other constrained matrix problems which arise in the estimation of O/D traffic flows and input/output matrices, has the sum of the elements of column i for all i, in the matrix. Hence, a formulation of the social accounting balancing problem is rarely applicable to other matrix estimation settings. Finally, I would like to add that jointly with A. Robinson, we can now handle general positive definite quadratic weighting matrices and additional linear constraints (besides the usual upper bound and transportation-type constraints) to enhance the modeling aspects.

Q.   Isn't it typical in financial markets that multipliers would depend on the flow? That is, if you interpret these multipliers as dividends, dividends will be flow dependent.

A.   Dividends are generally a certain fixed percentage of the value of a stock. Thus an arc multiplier models this situation quite properly. However, when multipliers are used for currency translations, the exchange rate, indeed, depends upon the amount of the transactions involved. Clearly arbitrage cannot go on forever before the effected prices change. You must set an upper bound on flow over arcs of this sort.

Q.   Is it possible, or even likely, that optimization models are widely used in the banking world but are simply not talked about?

A.   I don't see any evidence of this usage. Furthermore, I believe that wide recognition of financial planning models will not occur until the uncertainties are handled in an appropriate manner. Deterministic models have a limited range of application for planning purposes.

Q.   Many business transactions are done in fixed amounts. Do you have any procedures for handling integrality constraints?

A.   Due to objective function nonlinearities it is difficult to handle integer variables. Fortunately, the transactions discussed are quite large and heuristic approaches, such as rounding very small flows, seem to perform adequately. A decomposition approach would be an alternative strategy.

Q.   Explain the representation of flow for arcs connecting nodes in the network that occur in the future. When do these transactions take place?

A.   These transactions are carried out in the "current" future (or forward) market. A price is fixed today for an actual transaction that will take place at an agreed upon date in the future.

Q.   I would like to discuss my recent experience -- exchanging money. There are multipliers, certainly, also fixed charges. How are these handled?

A. Again, the size of the transactions involved overwhelms the fixed costs. Large corporations and banks are able to exchange money for an extremely low fee.

Q. In the context of the nonlinear algorithms, how do you carry out the preconditioner?

A. We use the diagonal of the reduced Hessian for preconditioning the conjugate gradient procedure.

Q. I am wondering about the numerical details behind the bonding procedure. Did you have any trouble with degeneracy?

A. The simplicial decomposition algorithm can be interpreted as a special case of Dantzig-Wolfe decomposition (with no complicating constraints). Thus the bonding mechanism does not depend upon non-degeneracy assumptions. In practice, the algorithm has been able to locate sub-optimal solutions (within .5% from optimality) in a short amount of computer time.

Q. What is the form of the objective function for the power transmission scheduling model?

A. There are several alternatives. A popular choice is to maximize the expected revenues produced by generating electricity. Remember that power generation depends upon both the amount of water and the head. Also, the shape of the reservoir becomes important. Combining these terms into a single nonlinear function is rather complex.

Q. What kind of stochastic program are you proposing for the financial model? Wouldn't a resulting model be huge?

C. The stochastic model could be solved practically as fast as the deterministic. If you only consider overage and underage, you should be able to come very close to the time it takes to solve the deterministic problem.

A. Yes, but the case of dependent random variables and stochastic multipliers requires much additional effort. Computers are becoming much cheaper and new algorithms, such as the multi-scenario approach of Rockafellar and Wets, are now available. These stochastic optimization problem will be a topic for much future research.

References

1. J. Abadie, "The GRG Method For Nonlinear Programming," in *Design and Implementation of Optimization Software*, ed. H.J.Greenberg, pp. 335-362, Sijthoff & Noordhoff, Netherlands, 1978.

2. P. G. Abrahamson, "A Nested Decomposition Approach for Solving Staircase Linear Programs," Report SOL 83-4, Department of Operations Research, Stanford University, 1983.

3. D. P. Ahlfeld, R. S. Dembo, J. M. Mulvey, and S. A. Zenios, "Nonlinear Programming on Generalized Networks," Report EES-85-7, to appear in *Transactions on Mathematical Software*, Princeton University, June 1985.

4. R.S. Barr, *The Multinational Cash Management Problem: A Generalized Network Approach*, University of Texas, Austin,Texas, 1972.

5. O. Barros and A. Weintraub, "Spatial Market Equilibrium Problems and Network Models," *Discrete Applied Mathematics*, vol. 13, pp. 109-130, 1986.

6. M.C. Biggs and M.A. Laughton, "Optimal Electric Power Scheduling: A Large Nonlinear Programming Test Problem Solved by Recursive Quadratic Programming," *Mathematical Programming*, vol. 13, pp. 167-182, 1977.

7. J. R. Birge, "Decomposition and Partitioning Methods for Multistage Stochastic Linear Programs," *Operations Research*, vol. 33, no. 5, pp. 989-1007, 1985.

8. J. Bisschop and A. Meeraus, "Matrix Augmentation and Partitioning in the Updating of the Basis Inverse," *Mathematical Programming*, vol. 13, pp. 241-254, 1977.

9. J. Bisschop and A. Meeraus, "On the Development of a General Algebraic Modeling System in a Strategic Planning Environment," *Mathematical Programming Study*, vol. 20, pp. 1-29, 1982.

10. G. G. Brown and R. D. McBride, "Solving Generalized Networks," *Management Science*, vol. 30, no. 12, Dec. 1984.

11. L. Cooper and L.J. LeBlanc, "Stochastic Transportation Problems and Other Network Related Convex Problems," *Naval Research Logistics Quarterly*, 1977.

12. R. L. Crum and D. L. Nye, "A Network Model of Insurance Company Cash Flow Management," *Mathematical Programming Study*, vol. 15, pp. 86-101, 1981.

13. G.B. Dantzig, *Linear Programming and Extensions*, Princeton University Press, Princeton, 1963.

14. R. Dembo, A. Chiarri, J. Gomez, L. Paradinas, and J. Guerra, "Integrated System for Electric Power System Operation Planning," Hidroelectrica Espanola Technical Report, Madrid, 1985.

15. M. Florian, "Nonlinear Cost Network Models in Transportation Analysis," *Mathematical Programming Study*, vol. 26, pp. 167-196, 1986.

16. R. H. Fourer, "Solving Staircase Linear Programs by the Simplex Method, 1 : Inversion," *Mathematical Programming*, vol. 23, pp. 274-313, 1982.

17. R. H. Fourer, "Solving Staircase Linear Programs by the Simplex Method, 2 : Pricing," *Mathematical Programming*, vol. 25, pp. 251-292, 1983.

18. R. Glassey, "Nested Decomposition and Multi-Stage Linear Programs," *Management Science*, vol. 20, no. 3, pp. 282-292, 1973.

19. F. Glover, J. Hultz, D. Klingman, and J. Stutz, "Generalized Networks: A Fundamental Computer-Based Planning Tool," *Management Science*, vol. 24, no. 12, pp. 1209-1220, 1978.

20. M.A. Hanscom, L. Lafond, L.S. Lasdon , and G. Pronovost, "Modeling and Resolution of the Deterministic Mid-term Energy Production Problem for the Hydro-Quebec System," Report IREQ-1453, L' Institut de Recherches de l' Hydro-Quebec, 1976.

21. H.H. Happ, "Optimal Power Dispatch - A Comprehensive Study," *IEEE Transactions PAS*, vol. 96, pp. 841-854, 1977.

22. P.T. Harker, "Alternative Models of Spatial Competition," *Operations Research*, vol. 34, No.3, pp. 410-436, May 1986.

23. D. W. Hearn, S. Lawphongpanich, and J. A. Ventura, "Restricted Simplicial Decomposition: Computation and Extensions," Research Report No. 84-38 ,(to appear *Mathematical Programming Study*), University of Florida, Gainesville, Oct. 1984.

24. J. K. Ho and A. S. Manne, "Nested Decomposition for Dynamic Models," *Mathematical Programming*, vol. 6, pp. 121-140, North-Holland, 1974.

25. B.F. Hobbs, "Network Models of Spatial Oligopoly with an Application to Deregulation of Electricity Generation," *Operations Research*, vol. 34, No.3, pp. 395-409, May 1986.

26. B. Von Hohenbalken, "A Finite Algorithm to Maximize Certain Pseudoconcave Functions on Polytopes," *Mathematical Programming*, vol. 13, pp. 189-206, 1975.

27. B. Von Hohenbalken, "Simplicial Decomposition in Nonlinear Programming Algorithms," *Mathematical Programming*, vol. 13, pp. 49-68, 1977.

28. C. A. Holloway, "An Extension of the Frank and Wolfe Method of Feasible Directions," *Mathematical Programming*, vol. 6, no. 1, pp. 14-27, 1974.

29. H. Markowitz, *Portfolio Selection*, Wiley, New York, 1959.

30. J. M. Mulvey, S. A. Zenios, and D. P. Ahlfeld, "Simplicial Decomposition for Convex Generalized Networks," Report EES-85-8, submitted to *Mathematical Programming* , Princeton University, 1985.

31. J. M. Mulvey and S. A. Zenios, "Real-Time Operational Planning for the U.S. Air-Traffic System," Research Report (submitted to *Applied Numerical Mathematics*), Princeton University, Princeton, New Jersey, 1986.

32. J. M. Mulvey, "Anticipatory Personnel Management: An Application of Multi- Criteria Networks," to appear in *Management of R & D Engineering* , North-Holland, 1986.

33. J. M. Mulvey, "Nonlinear Network Models in Finance," *Advances in Mathematical Programming and Financial Planning*, vol. 1, pp. 253-271, JAI Press, 1987.

34. B. A. Murtagh and M. A. Saunders, "MINOS User's Guide," Report SOL 77-9, Department of Operations Research, Stanford University, California, 1977.

35. L. Qi, "Forest Iteration Method for Stochastic Transportation Problem," *Mathematical Programming Study 25*, pp. 142-163, 1985.

36. R. T. Rockafellar and R. J-B. Wets, "Scenarios and Policy Aggregation in Optimization Under Uncertainty," WP-87-119, IIASA, December 1987.

37. J. B. Rosen, "Convex Partition Programming," in *Recent Advances in Mathematical Programming*, ed. P. Wolfe, McGraw-Hill, New York, 1963.

38. R. E. Rosenthal, "A Nonlinear Network Flow Algorithm for Maximization of Benefits in a Hydroelectric Power System," *Operations Research*, vol. 29, pp. 763-786, 1981.

39. A. Rudd and B. Rosenberg, "Realistic Portfolio Optimization," in *TIMS Studies in Management Sciences*, vol. 11, pp. 21-46, 1979.

40. D. P. Rutenberg, "Maneuvering Liquid Assets in a Multi-National Company : Formulation and Deterministic Solution Procedures," *Management Science*, vol. 16, no. 10, pp. B671-B684, 1970.

41. A.M. Sasson and H.M. Merrill, "Some Applications of Optimization Techniques to Power Systems Problems," *Proceedings IEEE*, vol. 62, pp. 959-972, 1974.

42. A.M. Sasson, W. Snyder, and M. Flam, "Comments on "Review of Load-Flow" Calculation Methods," *Proceedings IEEE*, vol. 63, pp. 712-713, 1975.

43. D. M. Scott, "A Dynamic Programming Approach to Time-Staged Convex Programs," Report SOL 85-3, Department of Operations Research, Stanford University, 1985.

44. S. W. Wallace, "A Piecewise Linear Upper Bound on the Network Recourse Function," CMI No. 862330-3, Department of Science and Technology, Christian Michelsen Institute, Bergen, Norway, February 1985.

45. S. W. Wallace, "Solving Stochastic Programs with Network Recourse," *Networks*, vol. 16, pp. 295-317, 1986.

46. R. J. Wittrock, "Advances in Nested Decomposition Algorithms for Solving Staircase Linear Programs," Report SOL 83-2, Department of Operations Research, Stanford University, 1983.

47. W. T. Ziemba, "Computational Algorithms for Convex Stochastic Programs with Simple Recourse," *Operations Research*, vol. 18, pp. 414-431, 1970.

Mathematical Programming as an Intellectual Activity

A.H.G. Rinnooy Kan
Econometric Institute
Erasmus University
P.O. Box 1738
3000 DR Rotterdam
The Netherlands

1. INTRODUCTION

For every scientific discipline, a process of reappraisal and evaluation of
the discipline as a whole forms a natural topic for informal discussion among
scientists. These discussions – typically held between lectures or after
dinner – might focus on the promise held out by a new development, on the
tenacity of certain venerable subareas within the discipline or on the growth
(or decline) of prominent research groups. In every case, the discussion is
part of a natural mechanism through which a scientific community strives for
some consensus on what the state of the discipline is and where it might be
heading next.

A precise discussion of this mechanism and of the results that it produces
will obviously vary among disciplines. Followed over the years, it leads to an
intellectual history of a discipline that may well reflect prolonged
differences of opinion and that will inevitably be influenced by many
subjective contributions to what is essentially a social process. None the
less, it is through this intellectual history that the actual development of
the discipline can best be understood if not predicted.

In what follows, an attempt will be made to outline the typical features of
this approach to history when applied to the field of mathematical
programming. By referring to mathematical programming as an intellectual
activity, we hope to indicate an attractive middle road between the abstract
and the anecdotal. This is further discussed in Section 2. There, it is also

NATO ASI Series, Vol. F51
Algorithms and Model Formulations
in Mathematical Programming
Edited by Stein W. Wallace
© Springer-Verlag Berlin Heidelberg 1989

argued that mathematical programming can best be viewed as a part of <u>applied</u> <u>mathematics</u>, and that the <u>dynamics</u> of the area and its subareas can be conveniently described along lines that are appropriate for the larger field as well.

In Section 3, as a very first – and very modest step – towards the description of mathematical programming as an intellectual activity, a classificatory analysis is presented of the set of papers presented at the thirteen mathematical programming symposia. To the extent that this analysis appears to confirm the existence of trends within the discipline, some tentative conclusions are drawn.

2. BACKGROUND

There are several ways to conceive of the history of a scientific discipline.

Of course, one important source of information consists of <u>textbooks</u> and <u>review articles</u>. In these, an effort is made to organize the body of available knowledge so as to emphasize essential scientific concepts and the relations between them. Generally, the authors will tend to focus on the issues that <u>in</u> <u>retrospect</u> turn out to have been most fruitful.

To the extent that this literature addresses the history of the discipline at all, it does so from a perspective that is totally dominated by current insights and preferences. This is not surprising: major and minor revolutions in science are usually accompanied by a rewrite of its history. A triumphant new regime will quickly produce an updated view of the scientific universe which then serves as a frame of dogmatic reference and an ideological source of problems until the next breakthrough occurs. Indeed, good textbooks are typically praised for their ability to reorganize a mass of historical material in such a way that it seems to lead up inescapably to today's priorities. Needless to say, in the process of tracing these developments many historical suggestions that appear to be misplaced are conveniently ignored, although they may well have to be revived in the future.

In every case, this <u>abstract</u> history never tells the full story. Scientists seem to be aware of this shortcoming, as they frequently supplement it by what could be called the <u>anecdotal</u> history. Typicallly, such anecdotes depict the grand old men of the discipline in their early struggle to gain recognition for new ideas. An example cherished by mathematical programmers is the story told by Dantzig about Hotelling's comment on his linear programming talk: 'But the world is nonlinear!' Von Neumann's intervention to the effect that someone living in a nonlinear world should just not use the method apparently saved the day for the profession.

Many such anecdotes exist, and they undoubtedly contribute to the social cohesion of the discipline. Some of them serve very well to illustrate the historical climate in which the discipline came to flourish, some just fondly recall the eccentricities of the scientific pioneers. But these fables again do not tell the full story. The main characters involved were not brought together randomly, but in the context of a purposeful scientific effort that motivated many of their actions.

In a description of mathematical programming as an intellectual activity, an attempt should be made to incorporate both sources of information. The research community itself would tend to view the area primarily as a set of ideas, but it is more than that: it also refers to a <u>locus of activity</u>, the perception of which varies strongly among individuals, a <u>social category</u> that should also be analyzed in terms of the collective biographies of the main actors. If this broader view of a scientific discipline is accepted, then it must also be accepted that the physical actions of those who erect the social category are important in determining how it evolves over time.

The above approach would recognize the existence of a discipline not only as a set of ideas that is extended and adapted as the result of new insights, but also as a collection of scientists who adopt a common professional label, and who are accepted under that label by neighboring disciplines. Thus, to determine the appropriate perspective from which to evaluate the development of mathematical programming, a natural start is to recall the disciplines that could be viewed as the <u>intellectual roots</u> of the profession.

The discipline of <u>mathematics</u> must rank high on that list. Indeed, setting first derivatives to zero is as close as most people get to mathematical programming. The challenge to come up with an appropriate mathematical treatment of more complex optimization problems continues to justify most of the academic research in the area, focused as it is on mathematical programming as an <u>abstract mathematical language</u>.

Of course, the mathematical interest in optimization is closely linked - since Newton's days - to its use in <u>physics</u> and other <u>exact sciences</u> as an <u>explanatory device</u>. The development of physical systems can be fruitfully analyzed in terms of its minimization of, e.g., a potential energy function. As such, optimization has been tremendously successful, and not surprisingly its use in this context has spread to the <u>social sciences</u>, most notably economics. Here, optimization serves as a <u>behaviorial paradigm</u>, e.g. in the assumption that consumers behave as if they maximize a utility function, subject to budget constraints.

Economists, however, have also noted the possible use of optimization as a <u>policy tool</u>, through which (supposedly) optimal policy decisions can be made. This aspect of mathematical programming is also at the heart of its use in <u>engineering</u>, where it serves to compute optimal designs or constructions. Similarly, <u>operations researchers</u> have been quick to notice the potential use of optimization as a <u>decision technique</u> for application in the context of an appropriate mathematical model.

The relation between operations research and mathematical programming is a particularly interesting one, since the former discipline might argue with some historical justification that it is the true godparent of the latter. Certainly, linear programming is the founding technique for both. As happens to other areas, operations research appears to be doomed to be an '<u>incubator science</u>', the promising parts of which break away quickly to grow into viable separate disciplines of their own. Certain subareas of mathematical programming currently exhibit similar tendencies.

The last intellectual root of mathematical programming that deserves mention is <u>numerical analysis</u>, the science of manipulating numbers through the use of machines. The connection to mathematical programming here is a particularly obvious one; it continues to be reflected in joint scientific meetings and workshops.

How will mathematical programming be regarded and according to which criteria would it be evaluated from these founding disciplines? Mathematicians, with the true arrogance of their profession, would judge it exclusively on its mathematical depth and elegance. They will have little appreciation for the practical usefulness that would form the primary criterion for engineers and management scientists. Computer scientists would look at the computational progress made and at the sophistication of the tools used. Different though these viewpoints are, a common denominator does appear to be the function of mathematical programming as a _methodological tool_ that is applicable within mathematical representations of the real world. Thus, it is appropriate to judge and describe the area as a specialization within _applied mathematics_. In doing so, one incurs the automatic disdain of some mathematicians, who, relying on Hardy's arguments in favor of the uselessness of mathematics, equate applied mathematics with bad mathematics. Mathematical programmers will just have to learn to live with this derision (and laugh all the way to the bank).

It was argued above that the fate of a discipline rests in the hands of those who actively recognize it as a separate social category. In the case of applied mathematics, this implies that the dynamics of any subarea are determined by the perceptions and actions of three groups:
- the _insiders_: the professional specialists for whom the subarea represents the dominant part of mathematics, coloring their perception of mathematics as a whole;
- the _colleagues_: the forum of (applied) mathematicians at large where the respectability of the area as a formative part of mathematics is being weighed;
- the _customers_: the real world in which the area is judged by its ability to contribute useful ideas for the solution of practical problems.

It is through the importance of this latter group that a fundamental difference arises between pure and applied mathematics. In pure mathematics, the evolution of an area is wholly internal to mathematics itself; in applied mathematics, an area can afford a lack of theoretical glamor provided that the real world offers compensatory appreciation.

Given the three relevant group of actors, it is possible to sketch the typical development over time of an area or subarea in applied mathematics.

In the beginning, at <u>birth</u>, there always is a <u>new opportunity</u>. To generate such opportunities is surprisingly easy: in principle, every insider or outsider is free to do so, although of course only a small fraction will lead to a substantial follow-up. Here, mathematics differs fundamentally from the exact and social sciences where new opportunities are usually associated with the inability of the discipline to account for certain empirical observations. The unusual relationship between new science and old science in mathematics will reappear in various guises below.

In mathematics, a new opportunity may correspond to a new question or a new tool; in applied mathematics, both can be theoretical or practical. In mathematical programming, e.g., the theory of NP-completeness provided a new theoretical tool and parallel computers a new practical tool.

In either case, the new opportunity has to be judged <u>relevant</u> by mathematical colleagues or customers (or, ideally, by both). If not, there will be little incentive to pursue it at length. Equally important though, the opportunity must appear to insiders to be a <u>manageable</u> one; there must be a glimpse of a possible answer, if only in the form of an endorsement by an acknowledged leader in the field whose intuition can be relied upon. A Dutch proverb states that any fool can keep ten wise people guessing: asking questions is easy enough, asking interesting questions is not difficult either, but asking promising questions is an ability not granted to everyone.

The new opportunity also has to be <u>competitive</u>; it has to be able to draw and hold attention from specialists who might be engaged in other work right now. Here again, mathematics differs from the empirical sciences: new theories do not invalidate old ones, but compete with them for the attention of a scientific community whose members are usually sufficiently flexible to be able to switch allegiance quickly. A new mathematical opportunity has to be <u>revolutionary</u> to attract supporters, not so much by destroying an old set of beliefs but by gently nudging them aside, inviting the spectators to consider the unparallelled opportunity for scientific (and worldly) rewards that the new opportunity is holding out. A few pioneers may then jump on the bandwagon, and a minor stampede might follow soon. In areas such as mathematical programming, researchers are always eager to cash in quickly when a promising opportunity arises; the juiciest fruits are usually picked by the first explorer to arrive on the scene.

If all goes well, the (sub)area may reach <u>maturity</u>, a phase that deserves to be characterized both in the <u>abstract domain</u> of ideas and in the <u>real domain</u> of human action and interaction.

In the abstract domain, then, a mature subarea will be characterized by a reasonably defined set of questions, a reasonably demarcated set of tools that can hopefully be used to answer these questions and a few initial, partial but encouraging answers. What remains are a larger number of <u>puzzles,</u> a term used by Kuhn to indicate well-defined technical questions to which the answers are unlikely to be shocking or revolutionary, but none the less of interest to the scientific subcommunity that has now been formed. The initial opportunity may well have been reformulated by now, some initial hunches may have been proved wrong, but the founding ideas are still recognizable and continue to inspire optimism.

In the real domain, researchers so inspired will by now have formed a <u>network</u> that provides for the informal exchange of information on scientific progress. Sometimes referred to as 'invisible colleges', these networks play an indispensable role in keeping the subcommunity informed about the forefront of research and, equally importantly, about disappointing forays. Increasingly supported by the electronic mail system, these networks precede official publications by such a long period that no serious researcher in the subarea can afford not to be included.

Then, of course, there are specialized meetings or specialized sessions at larger meetings, later often remembered nostalgically for their pioneering, enthusiastic atmosphere. Research funds are coming in, and the professional journals are beginning to take notice of the increasing stream of publications, e.g. by appointing members of the new subarea to their editorial boards. Perhaps most importantly, graduate students are motivated to settle down in the community and choose an appropriate puzzle as a topic for their Ph.D. research. Without the latter phenomenon no area can survive: Nachwuchs, scientific offspring, is essential for continuity.

Survival is, indeed, what the next phase is about. To consolidate its position, a new area has to continue to be attractive and to compete success-fully. In the abstract domain, the encouraging signs to look for are the emergence of new questions, either enlarging the intellectual domain or increasing its depth, and of new relations to other subareas or disciplines. These new relations may take the form of subsumption, i.e., when it can be demonstrated that the new approach subsumes earlier work as a special case, or of interfacing when fruitful questions can be posed on the borderline between disciplines. (the inspiring mixture of combinatorial optimization and probability theory is a good example of such an interface.) These expansions are not viewed as revolutionary; rather, the happily confirm the intuition of the pioneers that they have identified a topic of lasting interest. In applied mathematics, increasing success in the real world can compensate for the lack of spectacular theoretical developments; indeed, some would argue that this is where the current hopes for mathematical programming as a whole ought to lie.

In the real domain, survival is usually associated with further signs of recognition of the subarea as an important and relevant one. The subject matter will appear in standard university curricula, the first textbooks will be published and ad hoc university research groups may be replaced by official departments handing out specialized undergraduate degrees. Graduate students continue to be attracted (the first draft may have reached supervisory status by now), and research funding agencies no longer have to be reminded about the importance of the subarea. Recognition also ought to come from outside colleagues; if the subarea isolates itself too quickly, it could become a society for mutual admiration, and it may lose its credibility wihtin the larger community. Scientific prizes are an encouraging sign, not of course those created and awarded within the subcommunity itself, but those that come from a larger community, such as a Turing Award or the ultimate accolade of a Nobel Prize. Again, for applied mathematics, an increasing reputation among customers can compensate for a lack of scientific endurance; in the real domain, customers should be more than convinced that, in the words of Gertrude Stein, there is a useful 'there there', and express their appreciation by taking the initiative for large contracts rather than grudgingly assenting to small experiments.

So much for survival. Regrettably, not all subareas do survive forever and signs of <u>decline</u> may become visible at some point. In the abstract domain, decline manifests itself through an increasing conviction among insiders that the theoretical framework is exhausted: most fundamental questions seem to have been answered and the puzzles that remain appear to many to be routine challenges at best. Outsiders within the mathematical forum may already be aware of new ideas that compete effectively with the current ones in the subarea; the customers become aware of the fact that the practical effectiveness of the subarea is leveling off. It is important to recognize that the perception by insiders may differ dramatically from that by outsiders; the former tend to cling to the idea that has served them so well in the past and may continue to pursue the remaining puzzles with a ferocity that is only matched by the growing indifference demonstrated by colleagues and customers alike.

These developments are easily recognizable in the real domain. The number of PhD students will decrease, research grant applications are turned down, articles are increasingly rejected, and invited talks and special sessions become rare events. Also, technical notes start to replace substantial articles; they generally tidy up loose ends or present abstract generalizations of earlier results of no perceivable elegance. If the inner resistance within the subcommunity is sufficiently high, the specialized journals that once appeared to be a sign of external recognition may now start to play the role of the last stronghold: the community increasingly talks to itself, in isolated meetings without outside attendance. An atmosphere of intellectual provincialism may start to prevail. The original pioneers may not be around anymore; their above average intuition may have indicated better opportunities elsewhere, and with the flexibility that is typical for mathematicians they may already have jumped on another bandwagon taking their PhD students with them.

The end of the story need not be depressingly dramatic. A subarea may simply have found its proper niche within the framework of a larger scientific discipline. There it remains, as a standard item in a textbook, a standard tool for the practical problem solver, or a standard concept for the theoretical analyst. (NP-completeness and Lagrangean relaxation are good examples within mathematical programming.) If, however, earlier promises have not been realized at all, then the subarea may really fade away along with the remaining old soldiers that still loyally defend it, to be remembered only as a curiosity or as an interesting exercise for an historian of science.

Presumably, readers will have recognized some of the phenomena described above as having occurred within the area of mathematical programming. To investigate them in more detail and to document the dynamics of the area is the challenge that remains. A rational reconstruction of the history of the discipline, in the form of a description of its intellectual growth with due respect for the time dimension but with the speedometer set so as to bring out the logic of proofs and refutations (in Lakatos' terms), of questions and counterquestions, would then have to be written up. It should make use of the collective biographies of the important actors, written with due consideration for their environment (the Air Force, RAND), their careers, their students and their movements in time and space. That project would be altogether too big for this modest contribution; but a quick and dirty sample of the intellectual history of mathematical programming is offered in the next section, if only to wet the appetite.

3. TRENDS IN MATHEMATICAL PROGRAMMING

The notion that certain trends exist within mathematical programming so that its subareas expand and shrink over time, is an intuitively convincing one to most members of the community. Indeed, as new opportunities arise, the area as a whole attracts newcomers and at the same time the interests within the existing community may shift. Of course, people voting with their feet may well be obsessed by short term gains and ignore long term options; fifty years from now, the most useful work in retrospect may have been carried out by an isolated investigator slaving away on a homotopy method to solve nonlinear stochastic integer programming problems, without any official recognition in his (or her) lifetime.

With this valid objection against trend analysis in mind, one never the less wonders if there is any way in which the identification of changing scientific fashions over time can be given a more solid foundation than pure introspection would yield. To do so, it is tempting to focus on official publications such as journal articles. That, however, would seem to overemphasize the abstract history, and in addition would raise nearly insurmountable practical problems: mathematical programmers have published in such a broad range of journals that the boundary of the profession would have to be set either arbitrarily tight (thus, losing on representativity) or alarmingly broad (thus, losing on hours of sleep).

As a first step, it seemed much more attractive to examine the meetings with which the mathematical programming community as a whole identifies itself most explicitly. Fortunately, such a set of meetings exists in the form of the thirteen mathematical programming symposia that have been held between 1949 and 1985. A practical definition of a mathematical programmer might indeed be that everybody who attended at least one of those meetings is welcome to operate under that label. In using these meetings to analyze scientific trends, the assumption would be that they do not only reflect the overall size of the community and through that the respect in which the area is held, but also the relative interest attracted by various research topics over time. Again, both a decrease and an increase in this interest as measured by the number of lectures on the topic in question may well be misguided - but only with the benefit of hindsight.

Table 1 lists the symposia by year, and by location; it also contains the total number of lectures appearing in the official program.

Table 1

Mathematical Programming
Symposia

Index	Year	Location	Number of lectures
0	1949	Chicago	34
1	1951	Washington D.C.	19
2	1955	Washington D.C.	33
3	1959	Santa Monica	53
4	1962	Chicago	48
5	1964	London	74
6	1967	Princeton	87

Index	Year	Location	Number of lectures
7	1970	The Hague	138
8	1973	Stanford	198
9	1976	Budapest	303
10	1979	Montreal	471
11	1982	Bonn	503
12	1985	Cambridge	562

These programs still make fascinating reading, especially the earlier ones. The first (or rather, zeroth) symposium in 1949, for example, is remarkable for its star-studded cast of participants, with a particularly strong delegation of economists: Koopmans, Georgescu-Roegen, Dorfman, Simon, Morgenstern, Marschak and others. Where have all the economists gone? In a story told by Wolfe, these were the days that mathematical programmers did not want to be disturbed in their offices because they were building detailed models of the American economy! The economists may simply have become disillusioned.

Most talks during the first meetings were on the subject of solution methods and application of linear programming. The 1951 symposium was announced under the title 'Linear inequalities and programming' and was strongly influenced by the 'Scoop' project of the Air Force, featuring among other things the announcement that a special Scoop machine would be built to reduce the solution time of the planning model from six months to something more manageable. By the time of the 1955 symposium (still on 'Linear programming'), the simplex method had come out as a clear computational winner. For the first time, stochastic programming figured prominently on the program, and the traveling salesman problem also made its first appearance (in a paper by Heller). The 1959 symposium was on 'Linear programming and recent extensions', the latter ones including network models (with papers by Flood, Ford, Shapley and Minty) and nonlinear programming (with papers by Rosen, Zoutendijk, Houthakker and others).

From 1962 onwards, the symposium was officially devoted to 'Mathematical programming', and slowly all the subareas flourishing today made their appearance: integer programming in 1962, matroids in 1967, fixed point computation in 1973 and so on. The group of participants broadened, including more and more non-Americans (Beale, Kantorovich, Prekopa and others surfaced in London in 1964), until in 1970 the time was ripe for the foundation of the Mathematical Programming Society and its journal. Attendance increased steadily, with a quantum jump in Montreal where the number of lectures almost reached 500, a level that has been surpassed since then. Interestingly, most of the recent meetings continue to be memorable for certain developments in linear programming: Khachian in 1979, Borgward and Smale in 1982, Karmarkar in 1985. By extrapolation, one wonders what the 1988 symposium in Tokyo will bring!

Classifying the almost 3000 papers that were presented at the symposia was not a trivial task, and occasionally involved decisions to divide a paper among various categories. Readers inspecting the list in Table 2 will realize the inevitable overlaps that occur.

Table 2   Subareas of mathematical programming

LP	Linear programming
NLP	Nonlinear programming
IP	Integer programming
SP	Stochastic programming
CO	Combinatorial optimization
N	Networks
G	Game theory
C	Convex optimization
H	Homotopy methods
MC	Multicriteria methods
NDP	Nondifferentiable optimization
GO	Global optimization
A	Applications
EA	Economic applications
M	Miscellaneous

Integer programming was defined rather narrowly to refer to general (mixed) integer linear programming, i.e., not exploiting special structure, so as to attempt to distinguish it from combinatorial optimization. Networks was taken to include minimum nonlinear cost network flow models and its special cases. Convex optimization is also a problematic category; it could conceivably have been added to nonlinear optimization. Economic applications, as distinguished from applications are meant to include quasi-applied contributions to theoretical economics, in the sense these might not include documented computations in the real world. The category of miscellaneous includes papers on dynamic programming, control theory and related areas. While many of these choices are arbitrary, the overall picture probably would not change much as a result of different operationalizations.

The results are presented in Table 3, and given in the form of percentages rather than actual numbers.

Table 3

	0	1	2	3	4	5	6	7	8	9	10	11	12
LP	32	47	33	23	25	12	13	10	14	8	8	9	10
NLP	–	–	6	17	15	16	26	23	35	18	16	21	17
IP	–	–	–	8	10	10	11	17	17	9	8	3	5
SP	–	–	12	8	8	3	7	4	1	4	5	2	4
CO	–	–	3	2	2	8	8	10	5	17	12	27	21
N	–	–	–	17	8	7	7	2	5	8	6	4	6
G	4	5	–	–	–	4	2	3	1	3	2	2	2
C	3	5	–	9	6	7	2	2	3	5	6	4	1
H	–	–	–	–	–	–	–	–	2	1	3	1	2
MC	–	–	–	–	–	–	–	5	–	1	2	2	1
NDP	–	–	–	–	–	–	–	–	–	1	2	3	2
GO	–	–	–	–	–	–	–	–	–	1	2	2	4
A	18	32	26	8	15	15	7	14	11	11	15	8	12
EA	28	11	17	4	4	5	8	1	4	5	2	4	6
M	15	–	3	6	6	14	8	9	3	8	11	8	5

This disguises the fact that a two percent share among fifteen areas in 1985 is much more impressive in _absolute_ terms than a ten percent share in 1955. Readers eager to do so can easily compute the absolute numbers and be reassured that their favorite research area is not really becoming extinct.

One hopes that it is unnecessary to emphasize that any conclusions to be drawn from this material are hypothetical and subjective. Nevertheless, the temptation to offer some conclusions is irresistible, if only as an exercise in the gentle art of how to win enemies and influence people.

First of all, the analysis underlines the preeminence of _linear programming_ as the core technique of mathematical optimization. In retrospect, one can only admire the intuition of Dantzig, Hurwicz and other founding fathers for realizing the unique mixture of applicability and tractability that is offered by linear programming formulations. It is not hard to think of more widely applicable models (e.g., nonlinear integer programming) or of more tractable models, but a combination of both that is convincing to mathematical colleagues as well as to outside customers must be a rare event. As such, the research area of linear programming easily passes every survival test, up to the point that it continues to give rise to interesting new scientific challenges both related to the simplex method and to its competitors. It is hard to imagine a more impressive track record within applied mathematics.

Nonlinear programming in particular has benefited from the recent successful development of interior point methods for linear programming, in that it suddenly seems to offer a competitive approach to an area that was for a while considered to be the exclusive domain of the combinatorialists. In all the symposia, nonlinear programming emerges as a strongly represented research area, although progress appears to have been gradual rather than in spurts. Certainly, Karmarkar's work has provided a further impetus to this field.

The early promises of general _integer programming_ did not materialize, for computational reasons that are now much better understood than at the time. Techniques to solve general integer programs, i.e., those without special structure, have understandably not figured prominently on the program of recent symposia, as witnessed by Table 3. However, a revival may be on its way, in the wake of the succesful attempts by Padberg, Johnson and others to resurrect cutting plane approaches by exploiting ideas from polyhedral combinatorics.

Stochastic programming appears not to have delivered what it promised yet — but the expectations may have been unrealistically high. There is no doubt about the importance of mathematical programming under uncertainty from a modeling point of view, but the number of practical algorithmic success stories remains limited. The challenges are as interesting as ever, the difficulties perhaps to be alleviated by clever use of the new computational technology.

Combinatorial optimization has recently been astonishingly attractive to researchers. It does offer a good mix of tractability and applicability, but at the expense of an emphasis on special structure and hence of a loss of generality. The interaction with computer science and discreet mathematics has been a particularly stimulating one. One would suspect that the area is now so large and successful that subareas such as polyhedral combinatorics may well grow into viable subdisciplines of their own.

The success of specialized network algorithms has secured the continuing presence of this topic in all recent symposia, with the emergence of nonlinear network models in many applications suggesting an interesting new interface. Practical success here seems to be responsible for the prominence of this subarea at least as much as theoretical progress.

Once prominent areas such as convex optimization and game theory seem to be fading away, respectively to be subsumed under nonlinear optimization and to continue as a (regrettably) separate discipline. Homotopy methods appear not to live up to initial expectations. Multicriteria methods are increasingly studied in isolation from the main body of mathematical programming. Relatively young areas such as nondifferentiable and global optimization still have to establish their long term viability. The same is true for fashionable areas not appearing separately in the table, such as parallel computation and variational inequalities. One suspects that the former will exhibit spectacular growth in the near future.

Finally, but certainly not least importantly, <u>applications</u> are still part of
the scientific tradition represented at the meetings, but much less so than in
the early years. (This is particularly true for <u>economic applications</u>.)
Journal editors now find it hard to attract applied articles that are
realistic or truly innovative – let alone both. It may well be that many
successful applications are proprietary and hence never reported. Even so, one
is left with the impression that the process of <u>mathematization</u> and
<u>formalization</u> – in itself a sign of maturity – has led to the <u>internalization</u>
of most of the questions posed to the mathematical programming profession and
to the <u>externalization</u> of most of the answers. Put differently, the questions
coming from outside are manipulated until they fit within the conventional
wisdom of the current disciplinary framework, and the answers emerging from
the framework are not checked for external relevance but left at the doorstep
of the real world.

If this view is correct, then mathematical programming could end up by losing
the attractive mix of theoretical depth and practical usefulness that
characterizes the successful subareas of applied mathematics. The precarious
balance between the two aspects of the profession is worth preserving, and
since it is part of a social process, it requires a guardianship of good taste
that is the responsibility of the entire community.

The ideal balance should not be identified with the situation in which
algorithmic breakthroughs are primarily generated by questions from the real
world. That has rarely happened, although algorithmic details certainly have
been strongly influenced by what is demanded in practice. So far, the relation
between algorithms and modeling seems to have been mostly the other way round,
in that a model builder will tend to be guided by the (current) algorithmic
implications of the various possible mathematical representations of his
problem. To what extent this choice process can ever be elevated from an art
to a science remains to be seen.

Viewed as a whole, then, the discipline of mathematical programming would
appear to be a healthy one, not without its weaknesses, but with promises of
further progress (perhaps most strikingly in terms of exciting applications).
Such a discipline owes it to itself to preserve a careful history of its
intellectual past, if only by recording the views and recollections of its
founders. It is to be hoped that the mathematical programming society will
commission such an oral history before it is too late.

90

Acknowledgements

I am grateful to Philip Wolfe for supplying me with old symposium programs, and to Anneke van der Meer for her help in analyzing them.

The Aggregation Principle in
Scenario Analysis and Stochastic Optimization

Roger J-B Wets

Department of Mathematics
University of California
Davis, CA 95616

Research supported in part by grants from the National Science Foundation and the Air Force Office of Scientific Research.

NATO ASI Series, Vol. F51
Algorithms and Model Formulations
in Mathematical Programming
Edited by Stein W. Wallace
© Springer-Verlag Berlin Heidelberg 1989

1. Scenario Analysis

Most systems that need to be controlled or analyzed involve some level of uncertainty about the value to assign to some of the parameters, if not about the actual layout of certain subcomponents of the system. In many situations not much is lost by assigning "reasonable" values to these parameters or by choosing a particular design. In other instances ignoring uncertainty may very well lead to totally misleading solutions that would invalidate any of the implications one may wish to draw from the analysis.

Uncertainty can be included in the model in many ways. One that has been used successfully in a wide variety of situations is to model the uncertain quantities as random variables. In the case of decision-making problems, this leads us to *stochastic optimization models*, i.e., a problem that fits the following mold: given a probability space (Ξ, P), where Ξ is the set of possible realizations and P is the associated probability distribution, the problem is to

$$\text{minimize} \quad \mathbf{E}\{f(x, \xi)\} = \int f(x, \xi) dP(\xi)$$
$$\text{subject to} \quad x \in C \subset R^n,$$

where f is the criterion function and C is the set of feasible solutions determined by the constraints.

Because the model refers to ξ as a random vector with associated probability mass P, the model is not seen as an appropriate modeling tool when only limited information is available about the distribution of the random elements. In such cases, many practitioners often resort to *scenario analysis*. Uncertainty is to be modeled by a few *scenarios*, say

$$S = \{s^1, \ldots, s^L\},$$

and for each scenario $s \in S$, one finds the solution to the problem:

(1.1)
$$\text{minimize} \quad f(x, s)$$
$$\text{subject to} \quad x \in C_s \subset R^n.$$

Assuming the optimal solution exists for all s in S, let

(1.2)
$$x^s \in \text{argmin}\{f(x, s) \mid x \in C_s\}$$

denote such an optimal solution. We shall refer to it as a *scenario-solution*. Once scenario-solutions have been computed for all s in S, they are analyzed to see if general trends can be discovered, if there are clusters of solutions, how the solution x^s would "perform" if scenario s' is actually observed. An average solution is computed by assigning weights to the solutions x^s, and this "average" is then submitted to the same analysis, etc. The purpose of the analysis, of course, is to arrive at *one* solution that can be used for decision making. All the information gathered from the analysis will be used at arriving at a decision, but in all cases it always comes down to assigning relative weights to various factors, and then coming up with a "compromise". The guiding idea is that the solution selected should turn out to avoid disastrous situations whatever be the actual values taken on by the uncertain quantities, and in a certain average sense this solution should be optimal. Let

$$p_s, s \in S,$$

be weights that have been assigned to the (scenario-)solutions x^s to arrive at a decision x, viz.,

$$(1.3) \qquad \hat{x} := \Sigma_{s \in S} p_s x^s.$$

Because we are taking a combination of solutions, the p_s are necessarily nonnegative and add up to 1. These weights may have been available from the start (by relying on experts to assign relative importance to the various scenarios), or they may have been arrived at by the post-analysis sketched out above. It is not important at this stage to know exactly how these weights have been obtained. We can think of $\hat{x}$ as a "centroid" of the various scenario-solutions x^s, or simply as the *average* solution. In any case, this "compromise" solution can be viewed as "hedging" against uncertainty.

Although one may come up with very convincing arguments that a solution derived in this fashion is a "good" solution, in fact it is not a solution that takes into account the costs that may result from choosing decision x and observing scenario s (or, more precisely, events that are near scenario s). Since certain weights p_s are eventually going to lead us to the average solution $\hat{x}$, a solution that hedges against all eventualities would have been obtained by solving a problem of the type:

$$(1.4) \qquad \begin{aligned} &\text{minimize} \quad \Sigma p_s f(x, s^s) \\ &\text{subject to} \quad x \in \cap_s C_s, \end{aligned}$$

i.e., a stochastic optimization problem where scenario s is assigned probability p_s. Let us denote the optimal solution of this problem by x^*. The solution x^* does take into account the fact that it must "stand up" against a wide variety of possibilities.

However, practitioners of Operations Research continue to rely on scenario analysis for a number of reasons. The most common one is that problems of type (1.4) exceed our present computational capabilities. This sometimes ignores the progress that has been made in algorithmic development for stochastic programming problems, but it may be the case for very large (economic, energy policy, etc.) planning problems, exactly the type of problems that require taking uncertainty into account. Other reasons include (i) the fact that the individual scenario-solution x^s contains information that should help the decision maker, (ii) that the weights p_s being arbitrary, it should be possible to modify them to see what effect they would have on the solution and, (iii) that the use of parallel processors should allow us to process more than one scenario at a time and thus we could generate a "table" of solutions in not much more time than that required to generate just one scenario-solution.

Notwithstanding the fact that these reasons are sometimes well grounded, the major objection to the "scenario analysis" approach remains the lack of a solid and reliable mathematical basis for the justification of the solutions derived in this fashion. In this presentation, I will introduce a method that shows how to *aggregate* the scenario-solutions in an overall solution that will converge to the solution of the stochastic optimization problem (1.4). One of the advantages of this approach is that it does generate a "table" of individual scenario-solutions, and if intermediate solutions are analyzed, we can at an early stage identify the factors that play significant or insignificant roles in the construction of the overall solution. The aim of the algorithm is still to produce a solution of (1.4), but it only requires from us the capability to solve the individual scenario problems (1.1) and a simple perturbed version thereof. I shall also show how post-analysis of the weights p_s is possible without having to restart calculations from scratch.

This method, based on the *principle of scenario aggregation* in optimization under uncertainty, was developed in collaboration with Professor R.T. Rockafellar [8]. It can be interpreted as a decomposition-type method for stochastic programming problems (1.4). Indeed, we can think of the scenario problems (1.1) as subproblems of the stochastic programming problem (1.4), and the procedure is then to be viewed as one of "blending" the successive solutions generated by cost-modified versions of (1.1). In fact, this point of view leads to a new decomposition for stochastic programming problems that is not directly related to well-known primal and dual versions of the Dantzig-Wolf decomposition principle. In Section 6 we elaborate on this and show how this decomposition can be exploited in the environment of stochastic programming.

In Section 2 we restrict our attention to a simple scenario model, i.e., problems that involve a few scenarios and the sought-after solution is to be a straightforward aggregation of the scenario-solutions. A procedure, that could be named *sequential hedging*, is described and justified. Section 3 is a diversion on the *average problem*, a new concept that tries to characterize the problem to which $\hat{x}$ is an average solution. Section 4 extends the results to multistage problems, and it is seen how the concepts of implementability and feasibility enter into the search of an optimal solution. Section 6 is concerned with perturbation of the weights. And, as indicated earlier, Section 5 introduces a new decomposition for stochastic programming problems.

Although in this presentation I shall mostly restrict myself to the convex case, we shall see that in fact very few conditions need to be satisfied to make use of this procedure in a wide variety of situations. In fact, if one is interested only in finding a solution that improves on "pure" scenario analysis, then it should be possible to use the procedure in all situations that do not involve boolean (0-1) variables. And even then, running a few passes of the algorithm should be useful. The price to pay for such generality is that the procedure is probably not competitive (time-wise) with those that are aimed at solving linear, linear-quadratic, etc., stochastic programming problems. In fact, it should be clear from the way I have introduced the scenario's problem that the decision maker is confronted with a problem that could be formulated as a stochastic optimization problem, at least abstractly. But he does not really wish to treat it as a stochastic programming problem because the way he selected scenarios and "probabilities" would not even come close to satisfying any of the usual statistical criteria for defining probabilities. An advantage of this approach is that even if we do not pursue the search for an optimal solution to the very end, we always will have at hand an estimate of the optimal solution which is reliably better than any solution that could have been gleaned from "pure" scenario analysis.

2.   An Aggregation Algorithm

The method to be described here is certainly not the only one that could be viewed as an implementation of the principle of scenario aggregation. It starts with the solution to the scenario problems $\{x^s\} = \{x^1, \ldots, x^L\}$, to construct an estimate:

$$\hat{x} := \sum_{s \in S} p_s x^s,$$

that we have labeled the *average* solution. I shall refer to such a solution as *implementable*. What I mean by that, is that the decision $\hat{x}$ is scenario independent. One can view it as a decision that in the average (overall) will respond better to *all* eventualities than any particular scenario-solution x^s. (The full meaning of implementable will become clearer when we consider the multistage case in Section 4.) However $\hat{x}$ is not necessarily feasible.

A solution x is *admissible* if it is feasible for each particular scenario problem, i.e., if $x \in C_s$ for all s in S, or equivalently that $x \in \bigcap_{s \in S} C_s$. The solution of the stochastic optimization problem x^* is, of course, implementable and admissible. But in general, $\hat{x}$ will only be admissible if the sets of feasible solutions C_s are convex and they do not depend on s. But admissibility is not necessarily a *hard* constraint, in particular for problems of this type, and one could very well conceive that admissibility is a condition that could be waived, if the proposed solution was nearly admissible and possessed some other desirable characteristics. For example, if the violation only occurs for some unlikely scenarios, or if the violation is very mild, the decision maker may very well decide to accept the suggested solution. However, in our pursuit of an optimal decision, we shall seek a solution that is *feasible*, by which we mean that it is both implementable and admissible.

The particular realization of the principle of scenario aggregation, first described in [8], and reviewed here below, will generate a sequence of estimates $\{x^\nu, \nu = 1, \dots\}$ of the optimal solution x^*, obtained by insisting progressively more and more on the requirement that the solutions generated by the scenario problems (1.1) must be implementable.

For $\nu = 1, \dots$, let

$$\hat{x}^\nu = \sum_{s \in S} p_s x^{\nu s},$$

where

$$x^{\nu s} \in \operatorname{argmin}\{f^\nu(x, s) : x \in C_s\}$$

and

$$f^\nu(x, s) = f(x, s) + w^{\nu-1}(s)x + \tfrac{1}{2}\rho|x - \hat{x}^{\nu-1}|^2.$$

The vectors w^ν will be adjusted as indicated in Step 2 of the algorithm. We start with $\nu = 0$, and set

$$f^0(x, s) = f(x, s).$$

Algorithm.

Step 0. Initialize: set $w^0(s) = 0$, $\hat{x}^0 = 0$, choose $\rho > 0$ and set $\nu = 1$.

Step 1. For $s \in S$, compute $x^{\nu s}$.

Step 2. Calculate $\hat{x}^\nu$, return to Step 1 with $\nu = \nu + 1$, and

$$w^\nu(s) = w^{\nu-1}(s) + \rho[x^{\nu s} - \hat{x}^\nu].$$

Observe that

$$\sum_{s \in S} p_s w^\nu(s) = 0.$$

Indeed, from the definition of w^ν in Step 2 of the algorithm:

$$\sum_{s \in S} p_s[w^\nu(s) - w^{\nu-1}(s)] = \rho(\sum_{s \in S} p_s x^{\nu s} - \hat{x}^\nu) = 0, \ w^0(s) = 0.$$

Except for two simple calculations (updating w^ν and obtaining the estimate $\hat{x}^\nu$), the algorithm only requires the capability of solving individual scenario problems (and linear-quadratic perturbations thereof).

We have not yet discussed a terminating criterion. Since for all ν, the solution $\hat{x}^\nu$ obtained by taking a convex combination of scenario-solutions, is implementable, it will turn out that it is optimal as soon as $x^{\nu s} = \hat{x}^\nu$ for all s. To see this, observe that in such a case $\hat{x}^\nu$ is clearly feasible (it belongs to C_s for all s in S), and since for all $x \in S$,

$$\hat{x}^\nu \in \operatorname{argmin}\{p_s[f(x, s) + w^\nu(s)x + \tfrac{\rho}{2}|x - x^\nu|^2] : x \in C_s\},$$

it follows that

$$\hat{x}^\nu \in \operatorname{argmin}\left\{\sum_{s \in S} p_s[f(x, s) + \tfrac{\rho}{2}|x - x^\nu|^2] : x \in \bigcap_{s \in S} C_s\right\}$$

using here the fact that $\Sigma_s p_s w^\nu(s) = 0$. Assuming that the functions $f(\cdot, s)$ are "nice" enough in a neighborhood of $\hat{x}^\nu$, this in turn implies

$$\hat{x}^\nu \in \text{local-argmin} \left\{\sum_{s \in S} p_s f(x, s) : x \in \bigcap_{s \in S} C_s\right\},$$

and, in particular in the convex case, this tells us that $\hat{x}^\nu$ actually solves the stochastic optimization program (1.4). But even if the functions are nonconvex, we can still claim that $\hat{x}^\nu$ is a global minimum provided the functions $f(\cdot, s)$ satisfy certain minimum growth conditions.

At the conclusion of each major iteration, we can use the distance between the vectors

$$x^\nu = (x^{\nu s_1}, x^{\nu s_2}, \ldots, x^{\nu s_L})$$

and

$$\hat{x}^\nu = (x^\nu, x^\nu, \ldots, x^\nu),$$

as a measure of how close we are from satisfying all the constraints. This distance

$$\sum_{s \in S} p_s |x^{\nu s} - \hat{x}^\nu|^2 := \theta_\nu,$$

which can be interpreted as the conditional variance of the error relative to the weights p_s will converge to 0, at least in the convex case. That follows from the convergence results reviewed in Section 4.

3.   The Average Problem

For each ν, and each s, $x^{\nu s}$ is the optimal solution of the modified optimization problem

$$(3.1) \qquad \text{minimize } f(x,s) + w^{\nu-1}(s)x + \tfrac{\rho}{2}|x - \hat{x}^{\nu-1}|^2 + \psi_{C_s}(x)$$

where ψ_D is the indicator function of the set D with $\psi_D(x) = 0$ if $x \in D$, and $= +\infty$ otherwise. Excluding unusual circumstances, the average of the $\{x^{\nu s}, s \in S\}$, $\hat{x}^{\nu} = \sum_{s \in S} p_s x^{\nu s}$ is not the optimal solution of

$$\text{minimize } \sum_{s \in S} p_s[f(x,s) + w^{\nu-1}(s)x + \tfrac{\rho}{2}|x - \hat{x}^{\nu-1}|^2]$$
$$\text{subject to } x \in \bigcap_{s \in S} C_s$$

which in view of the fact that $E\{\mathbf{w}^{\nu-1}\} = 0$, can also be written as

$$(3.2) \qquad \text{minimize } E\{f(x,\mathbf{s})\} + \tfrac{\rho}{2}|x - \hat{x}^{\nu-1}|^2$$
$$\text{subject to } x \in \bigcap_{s \in S} C_s,$$

i.e., just a version of (1.4) with a penalty term. However $\hat{x}$, that we called an average solution, is the solution of a problem related to the scenario subproblems (3.1) that we now bring to the fore. The objective function of such a problem is not obtained by taking weighted sums of the (essential) objectives of (3.1), as we did when passing from (3.1) to (3.2), but by taking *epigraphical sums*. This operation is dual, in a sense to be made precise later on, to the (standard) addition of functions.

The *epi-sum*, or epigraphical sum, of two extended real-valued functions f and g defined on $\mathbb{R}^n$, is the function $f \underset{e}{+} g$ defined by:

$$(f \underset{e}{+} g)(x) := \inf_{u^1 + u^2 = x} [f(u^1) + g(u^2)].$$

If we write $\text{epi}\, h := \{(x, \alpha) \in \mathbb{R}^n \times \mathbb{R} \,|\, \alpha \ge h(x)\}$ for the epigraph of h, as a direct consequence of the definition, we have that

$$\text{epi}(f \underset{e}{+} g) = \text{vert-cl}\,[\text{epi}\, f + \text{epi}\, g]$$

where for a set $D \subset \mathbb{R}^n \times \mathbb{R}$, vert-cl D is the *vertical closure* of D, i.e.

$$\text{vert-cl}\, D := \{(x, \alpha) \,|\, \alpha \ge \inf[\alpha' : (x, alpha') \in D]\}.$$

Indeed

$$\text{vert-cl } [\text{epi } f + \text{epi } g]$$
$$= \text{vert-cl } \{(u^1 + u^2, \alpha_1 + \alpha_2) \mid \alpha_1 \geq f(u^1), \alpha_2 \geq f(u^2)\}$$
$$= \{(x, \alpha) \mid x = u^1 + u^2, \alpha \geq \inf[f(u^1) + f(u^2)]\}$$
$$= \text{epi}(f \underset{e}{+} g).$$

More generally, if $\{f(\cdot, s), s \in S\}$ is a finite collection of functions, and the $\{p_s, s \in S\}$ are probability weights, the *epigraphical mean*, or *epi-average*, of this collection is defined by

$$\left[e\text{-} \sum_{s \in S} p_s f(\cdot, s) \right] (x) = \inf \left[\sum_{s \in S} p_s f(u^s, s) \mid \sum_{s \in S} p_s u^s = x \right].$$

It is also convenient to write

$$(e\text{-}Ef)(x) \quad \text{for} \quad e\text{-} \sum_{s \in S} p_s f(\cdot, s);$$

the weights $\{p_s, s \in S\}$ are implicit in this notation. (In the literature the term *inf-convolution* has also been used to designate epi-sums and epi-means, however the latter terminology has important geometric content and should be preferred.)

One verifies easily that if the elements of the epi-mean, i.e., the $\{f(\cdot, s) s \in S\}$, are proper, lower semicontinuous functions, then

$$(e\text{-}Ef)^* = \sum_{s \in S} p_s f^*(\cdot, s)$$

where h^* is the conjugate of h,

$$h^*(v) = \sup[vx - h(x) \mid x \in \mathbb{R}^n].$$

Moreover if the $\{f(\cdot, s) s \in S\}$ are also convex, then $(e\text{-}Ef)$ is convex (see the formula for the epigraph of an epi-sum), and

$$\left[\sum_{s \in S} p_s f(\cdot, s) \right]^* = \text{cl}(e\text{-}Ef^*).$$

Finally, if the functions $\{f(\cdot, s), s \in S\}$ are inf-compact so is their epi-mean.

It is possible to extend the definition of epi-sum and epi-mean to arbitrary collections of functions. For example, if $(S, \mathcal{S}, P)$ is a probability triple, and $\{f(\cdot, s), s \in S\}$ any collection of functions, then the epi-mean is defined by

$$(e\text{-}Ef)(x) = \inf_{u \in \mathcal{M}} \left\{ \int_S f(u(s), s) P(ds) \mid \int_S u(s) P(ds) = x \right\}$$

where $\mathcal{M} :=$ {measurable functions from S into $\mathbb{R}^n$}. In the remainder of this section we shall adopt this more general definition of the average problem. When P is a measure with finite support, we are in the case considered in Sections 1 and 2, i.e., a finite number of scenarios.

Given a collection of optimization problems, parameterized by a finite or infinite collection of scenarios, to which are associated (probabilistic) weights, the *average problem*, is the epi-mean of the collection. The most important property of the average problem is the following:

Theorem. *Suppose that $(S, \mathcal{S}, P)$ is a probability space and f is an extended real-valued convex function defined on $\mathbb{R}^n \times S$, lower semicontinuous with respect to x (on $\mathbb{R}^n$) and measurable with respect to s (on S). Suppose also that for all s in S,*

$$\text{argmin} \, f(\cdot, s) := \Gamma(s) \text{ is a nonempty (convex) set.}$$

Then for any summable selection $x(\cdot)$ of Γ, we have:

$$\int x(s)P(ds) \in \text{argmin} \, e\text{-}Ef,$$

i.e., the solutions' average is the solution of the average problem (the minimization problem obtained by taking the epi-mean).

Proof. Let $x(\cdot)$ be such a summable selection, and $x^* = \int x(s)P(ds)$. From the definition of the epi-mean it follows that

$$e\text{-}Ef(x^*) = \int f(x(s), s)P(ds).$$

Moreover, for any other summable selection, say $x'(\cdot)$, for all s

$$f(x(s), s) \leq f(x'(s), s).$$

Since we are integrating with respect to a nonnegative measure, this implies that

$$e\text{-}Ef(x^*) \leq \int f(x'(s), s)P(ds).$$

Now, for any x in R^n, the value of $e\text{-}Ef(x)$ is obtained by choosing a summable $x'(\cdot)$ that satisfies $x = \int x'(s)P(ds)$, the above implies that

$$e\text{-}Ef(x^*) \leq e\text{-}Ef(x), \quad \text{for all } x \text{ in } R^n. \qquad \square$$

In general, it is not so easy to find a simple expression for the epi-sum of a collection. The formula given above, in terms of the conjugate functions, can be used in some instances, but explicit calculation of the conjugates could already be a major task. Whenever it is possible to find an expression for the average problem, it turns out, as can be expected, to be a problem with "smoother" properties. This is usually what occurs when averaging is involved.

In order to give the reader at least one example of an epi-sum, we now calculate the (simple) epi-sum of the two following functions:

$$f(x,1) := \tfrac{1}{2}\|x\|^2$$
$$f(x,2) := \|x\|$$

Then $(e\text{-}f)(\cdot) = f(\cdot,1) + f(\cdot,2) = [(\tfrac{1}{2}\|\cdot\|^2)^* + \|\cdot\|^*]^*$. Now $\|\cdot\|^{2*} = \tfrac{1}{4}\|\cdot\|^2$ and $\delta_B = \|\cdot\|^*$, where δ is the indicator function of B, the euclidean ball of radius 1. Hence $(e\text{-}f) = (\|\cdot\|^2/4 + \delta_B)^*$. A simple calculation yields

$$(e\text{-}f)(x) = \begin{cases} \tfrac{1}{2}\|x\|^2 & \text{if } x \in B; \\ \|x\| & \text{otherwise.} \end{cases}$$

(For this very simple example, one could also have obtained the epi-sum by a direct calculation.)

Another way to look at the average problem is to view it as a projection. Consider the (multi)function $s \mapsto \text{epi } f(\cdot,s)$ defined on S values in the epigraphical subsets of R^{n+1}. Then the epigraph of the average problem is the projection of this (multi)function on R^{n+1} under the map:

$$(x(\cdot), \alpha(\cdot)) \mapsto \left(\int x(s)P(ds), \int \alpha(s)P(ds) \right).$$

This point of view is important when considering the multistage model.

We now return to the algorithmic procedure introduced in Section 2. It is clear that at each iteration x^ν is the solution of an average problem obtained by calculating the epi-mean of the collection $\{f^\nu(\cdot,s), \ s \in S\}$ using the weights $\{p_s, \ s \in S\}$. Since the $f^\nu(\cdot,s)$ and $f^{\nu+1}(\cdot,s)$ differ only in the perturbation term that affect the objectives, it may be possible to take advantage of that feature to calculate the iterates x^ν. How to do this is an open question.

4. Multi-period Problems

The passage from single period problems to multi-period problems requires more than a simple enlargement of the problem's size (with the usual exploitation of structural properties). In optimization under uncertainty, it is also necessary to specify an information structure that models the information gathering process: *what* does the decision maker know and *when* does he know it.

Although decision stages do not necessarily correspond to time periods, let us index the decision stages by a time variable $t = 1, \ldots, T$, and write

$$x := (x_1, \ldots, x_T) \in \mathbb{R}^n \times \ldots \times \mathbb{R}^n$$

where the component x_t represents the decision to be taken at time t. More specifically, X denotes a function that assigns to each $s \in S$ a vector

$$X(s) := (X_1(s), \ldots, X_T(s)),$$

where $X_t(s)$ denotes the decision to be made in period t if we are confronted with scenario s (or by some event that is "near" scenario s). We call such a mapping a *policy*.

There is one important constraint on the choice of policies that is implicit in the formulation of the problem: *if two different scenarios s and s' are identical up to time t on the basis of the information available about them at time t, then up to time t the decisions $(X_1(s), \ldots, X_t(s))$ and $(X_1(s'), \ldots, X_t(s'))$ generated by the policy X must also be identical.* This condition guarantees that the solutions provided by the model do not depend (at time t) on information that is not yet available. If scenarios s and s' are identical up to time t, it means that the decision maker has no way to distinguish between these two eventualities, and his policy cannot assume that he knows which one of these two scenarios he is following.

Let $\mathcal{P}_t$ be the coarsest partition of S in sets, such that if $A \in \mathcal{P}_t$, and $s, s' \in A$, the scenarios s and s' are then undistinguishable up to time t. Mathematically, the condition that the decision at time t can only depend on the information available at time t, can be formulated as follows:

$$X_t(\cdot) \text{ must be constant on } A \text{ for each } A \text{ in } \mathcal{P}_t.$$

Let us stress the fact that in the modeling of the information process, it is important to give different labels to two scenarios that may have the same physical realization (outcome) but with different levels of information to be made available at time t. For example, a difference in two scenarios could simply be that in period t you will be given some prediction about the demand (market,...) at a later time period.

The *implementable policies* determine the subspace $\mathcal{N}$ of functions from S into $\mathbb{R}^{n \cdot T}$ that satisfy the preceding condition, i.e.,

$$\mathcal{N} = \{X : S \to \mathbb{R}^{n \cdot T} \mid X_t \text{ is constant on each } A \text{ in } \mathcal{P}_t\}.$$

If information is increasing with time, as it usually is, then $\mathcal{P}_{t+1}$ is a refinement of $\mathcal{P}_t$. But, neither does that need to be the case in practice, nor is that property necessary for the ensuing discussion. The multistage problem can thus be formulated as:

(4.1)   find a policy $x : S \to \mathbb{R}^{n \cdot T}$ that

$$\text{minimizes } \sum_{s \in S} p_s f(X(s), s)$$

subject to $X(s) \in C_s$ for all $s \in S$.

A policy is *feasible* if it is both implementable and admissible, where *admissible* means that

$$X \in \mathcal{C} := \{\text{policies that satisfy the (explicit) constraints}\}$$

$$= \{X \mid X(s) \in C_s \text{ for all } s \in S\}.$$

Policies that fail to be admissible may nonetheless be acceptable approximate solutions (contingency plans), either because the violated constraints are "soft" constraints, or the transgressions are only minor, or still, the violation occurs only for scenarios that are quite unlikely. As already indicated in Section 2, implementability is an inescapable requirement. And, this is reflected in the procedure that was introduced in Section 2 (for the 1-period problem) and of which I now give the full multiperiod version. The $\hat{x}^\nu$ calculated in Step 2 of the Algorithm in Section 2 are implementable solutions (that may fail to be admissible). They do not depend on s. For the multiperiod problem, the operation required to obtain an implementable policy from the solutions of the individual (perturbed) scenario problems, is somewhat more involved as taking a simple average with respect to the weights p_s. Since policies at any time t may depend on the information available, implementable policies are obtained by taking averages but conditioned on the information (that will become) available in each time period. For each $A \in \mathcal{P}_t$, let

$$p_A := \sum_{s \in A} p_s,$$

and define

$$\hat{X}_t(A) := \sum_{s \in A} p_s X_t(s) / p_A.$$

This vector represents the weighted average of all responses $X_t(s)$ to each individual scenario in "atom" A. An implementable policy $\hat{X} = (\hat{X}_1, \ldots, \hat{X}_T) \in \mathcal{N}$ is obtained by setting

$$\hat{X}_t(s) := \hat{X}_t(A) \text{ for all } s \text{ in } A.$$

The transformation $J : X \to \hat{X}$ is a projection (it is linear and $J = J^2$) that depends on the selected weights p_s. It will be called the *aggregation* operator relative to the given information structure $(\mathcal{P}_t, \; t = 1, \ldots, T)$ and the weights $(p_s, \; s \in S)$. If we interpret the weights as probabilities, then $\hat{X}_t$ is the conditional expectation of X_t given the sigma-field of events $\mathcal{P}_t$.

The *scenario aggregation principle*, which allows us to blend the solutions of individual scenario problems to generate a solution that is both implementable and admissible (and optimal in terms of problem (4.1)), provides the theoretical guidelines for the algorithmic procedure below. Once again, I want to emphasize that the particular algorithm given here is not the only procedure that could be generated by relying on the scenario aggregation principle to blend the solutions of the (individual) scenario problems.

Progressive hedging algorithm [8].

Step 0. Initialize: set $W^0(s) = 0$, choose $\hat{X}^0$, $\rho > 0$ and set $\nu = 1$.

Step 1. For each $s \in S$, solve (approximately) the optimization problem

$$\text{minimize } f^\nu(x, s) \text{ subject to } x \in C_s,$$

where

$$f^\nu(x, s) := f(x, s) + \sum_t [W_t^{\nu-1}(s) x_t + \tfrac{1}{2}\rho(x_t - \hat{X}_t^{\nu-1}(s))]$$

Let $X^\nu(s) := (X_1^\nu(s), \ldots, X_T^\nu(s))$ denote the vector of solutions.

Step 2. Update the perturbation term

$$W_t^\nu(s) := W_t^{\nu-1}(s) + \rho[X_t^\nu(s) - \hat{X}_t^\nu(s)]$$

where the "averaged" solution $\hat{X}^\nu$ is calculated as follows: for all s in $A(\in \mathcal{P}_t)$

$$\hat{X}_t^\nu(s) := \sum_{s \in A} p_s X_t^\nu(s)/p_A := E\{\mathbf{X}_t^\nu \,|\, A\}(s).$$

Return to Step 1 with $\nu = \nu + 1$.

Before we discuss the convergence properties of this algorithm, let us observe that for all ν, the vector W^ν is orthogonal to the subspace of implementable policies, for all $X \in \mathcal{N}$,

$$\sum_{t=1}^{T} \sum_{s \in S} p_s W_t^\nu(s) \cdot X_t(s) = 0.$$

In a probabilistic framework, this condition could be equivalently stated as $E\{W_t^\nu \mid \mathcal{P}_t\} = 0$. The algorithm is such that this condition is satisfied for each iterate. Indeed, it is trivially satisfied for W^0. For $\nu \geq 1$, if $s \in A$ (in $\mathcal{P}_t$)

$$E\{\mathbf{W}_t^\nu \mid s \in A\} = E\{\mathbf{W}_t^{\nu-1} \mid s \in A\} + \rho[E\{\mathbf{X}_t^\nu \mid s \in A\} - \hat{X}_t^\nu(s)]$$

where the first term of the right hand side is 0 by induction and the second term is 0 by definition of X_t^ν. There is nothing sacred about starting with $W^0(s) = 0$. In fact, we could start with any vector W^0 provided it satisfied the conditions: for all t, and all $A \in \mathcal{P}_t$

$$\sum_{s \in A} p_s W_t^0(s) = 0.$$

The resulting updates W^ν would again satisfy the conditions mentioned above.

The "multipliers" W^ν are themselves estimates of the price systems that can be associated with the (implicit) constraint that (feasible) policies must be implementable. We have expressed this by requiring that X should be in $\mathcal{N}$. An equivalent way to formulate this constraint is to require that X lies in the orthogonal complement of $\mathcal{N}^\perp$, i.e., the subspace orthogonal to that spanned by feasible W^ν. This can be expressed as: for all t, $A \in \mathcal{P}_t$, and $s \in A$

$$X_t(s) - \sum_{s' \in A} p_{s'} X_t(s') = 0.$$

The iterates of the W^ν can now be interpreted as estimates for the multipliers to be associated with these constraints. When the optimum is reached, let us say for a pair (X^*, W^*), there is a rich meaning that can be attached to the vector W^*. It can be viewed as an (equilibrium) price system to be associated with the lack of information. If information could be had (or if decisions could be adjusted), and the cost of this information (or adjustments) would be determined by the price system W^*, then the decision maker would have no incentives in buying this information (or buying the right to make a posteriori adjustments). For more about the interpretation to give to these multipliers, see Rockafellar and Wets [6], [7], Evstigneev [3], Back and Pliska [1], and Dempster [2].

Let us now turn to the justification of the algorithm and its convergence properties. For the reader that is familiar with augmented lagrangian methods, he may be misled by the air of similarity between that class of methods and the Progressive Hedging Algorithm. To more clearly identify the differences, let us briefly review lagrangian-based methods. We set

$$F(X) := \sum_{s \in S} p_s f(X(s), s),$$

and denote by $\langle X, W \rangle$ the inner product of X and W:

$$\langle X, W \rangle = \sum_{t=1}^{T} \sum_{s \in S} p_s W_t(s) \cdot X_t(s).$$

Problem (4.1) can be reformulated as:

(4.2)   minimize $F(X)$ subject to $X \in \mathcal{C}$, $X \in \mathcal{N}$

where $X : S \to \mathbb{R}^{nT}$, and $\mathcal{N}$ is determined by the linear constraints:

$$X_t(s) - \sum_{s \in A} p_s X_t(s) = 0, \text{ for all } t, \text{ and } A \text{ in } \mathcal{P}_t.$$

The associated lagrangian (where it is finite) is given by

$$L(X, W) = F(X) + \langle X, W \rangle \text{ if } X \in \mathcal{C}, \ W \in \mathcal{N}^\perp.$$

Let us presume that there is an overall strategy for improving iteratively the values to assign to the (dual) variables W. For fixed W° the problem to be solved is *decomposable*. We need to find $X^\circ(\cdot)$ that minimizes $F(X) + \langle X, W^\circ \rangle$.

This can be done by finding for all s, $X^\circ(s)$ that minimizes

$$f(X, s) + W^\circ(s) \cdot X \text{ subject to } X \in C_s.$$

We refer to this type of decomposability as *vertical decomposability* (by opposition to "horizontal decomposability" which here would mean decomposability with respect to time or stages). This is exactly what we are looking for, since it would mean that the difficulty in solving our problem is reduced to solving problems that are just simple perturbed versions of the original scenario problems. The difficulty in the implementation comes from the need to have an overall scheme that guides the adjustments that must be made to the iterates of W. This W-problem would be of the same size as the (full) original problem: the variable W is of dimension $n \cdot T$. And although we might be dealing with a problem with a simpler structure, we have not really made much progress in cutting down the size of the problem that needs to be solved to find an optimal solution of (4.1). If we were going to use this approach, we could, for example, rely on a cutting plane-type algorithm to solve the W-problem:

$$\text{maximize } \{G(W) = \inf_X L(X, W)\}.$$

Without going into the details, it is not clear how in this case the iterates provided by the solutions of the (decomposed) X-problem could be used to provide approximating solutions. And, if we did come up with some estimates for the solution, no use is made of these estimates in the algorithmic procedure.

A second possibility would be to use the augmented lagrangian method to provide the basis of the solution procedure, for example the *method of multipliers*. The augmented lagrangian (where it is finite) is defined by:

$$L_\rho(X, W) = F(X) + \langle X, W \rangle + 2^{-1}\rho\|X - \hat{X}\|^2, \hat{X} = \sum_{s \in S} p_s X(s) \text{ if } X \in \mathcal{C}, \ W \in \mathcal{N}^\perp,$$

$$= L(X, W) + 2^{-1}\rho\|X - \hat{X}\|^2.$$

The iterates are generated by:

$$X^{\nu+1} \in \operatorname{argmin} L_\rho(X, W^\nu),$$

$$W^{\nu+1}(s) = W^\nu(s) + \rho \left[X^{\nu+1}(s) - \sum_{s \in S} p_s \hat{X}^{\nu+1}(s) \right], \ s \in S.$$

The updates of the W-vector are easy enough to calculate, but the problem now is with generating the X-iterates. No longer can we rely on the vertical decomposability of the problem to reduce it to one which is not more involved than solving a collection of scenario-problems.

The implementation of the scenario aggregation principle that was first proposed in [8] avoids the pitfalls mentioned above. As we see from Step 2 of the Progressive Hedging Algorithm the updating of the W's does not require the solution of a high dimensionality problem and the present estimate is used in the procedure to stabilize the path followed by the X-iterates. And finally, it does decompose vertically which means that the complexity of solving this problem will never be more than that of solving individual scenario problems. It also provides the key to the use of parallel processing, bunching methods (that are "standard" in stochastic programming [9]), and other time saving procedures.

Convergence to a solution that is (implementable, feasible and) optimal for problem (4.1) can be proved under a wide range of assumptions. Please refer to [8] for convergence arguments that apply to linear-quadratic models, convex and non-convex models, and allow for approximate minimization. The result stated here covers only the convex case and assumes that the minimization in Step 1 of the algorithm is carried out exactly. Without explicitly mentioning it, we presuppose that the given problem is well-formulated and admits an optimal solution, somewhat more precisely: the lagrangian L is assumed to have a saddle point.

Theorem. [8, Theorem 5.1] *Suppose that for all s in S the objective function $f(\cdot, s)$ and the constraint-set C_s are convex, and that the problem satisfies the following constraint qualification:*

$$\{W \in \mathcal{N}^\perp \mid -W(s) \text{ is normal to } C_s \text{ at } X^*(s)\} = \{0\}$$

where X^* is the unique optimal solution. Then the X^ν generated by the Progressive Hedging Algorithm converge to X^*, and the W^ν converge to W^* with (X^*, W^*) a saddle point of the lagrangian L. Furthermore, the sequence $\{X^\nu, \nu = 1, \ldots\}$ provides a strictly improving sequence of estimates of X^* in the following sense:

$$\|(X^{\nu+1}, W^{\nu+1}) - (X^*, W^*)\|_\rho \leq \|(X^\nu, W^\nu) - (X^*, W^*)\|_\rho,$$

with strict inequality if the pair (X^ν, W^ν) is not optimal. Here

$$\|(X, W)\|_\rho = (\|X\|^2 + \rho^{-2}\|W\|^2)^{1/2}.$$

This suggests a linear convergence rate, at least in the (X, W)-space. This behavior is confirmed by the numerical experience that has been gathered so far [4]. The two figures below were graciously provided by John Mulvey (of Princeton University). They plot the quantity

$$\beta_\nu = \|X^{\nu+1} - X^\nu\| \text{ versus the iteration count } \nu.$$

We first see a rapid decrease in this quantity, followed by a first hump during which the procedure is adjusting the W-variables. This process is then repeated, again first a rapid decrease in the measure of the distance between the X-variables followed by a few iterations where the W-variables are seen to be of prime concern. Figure II is a rescaling to enlarge the third hump that appears in Figure I.

CONVERGENCE PERFORMANCE

OF SCENARIO AGGREGATION ALGORITHM

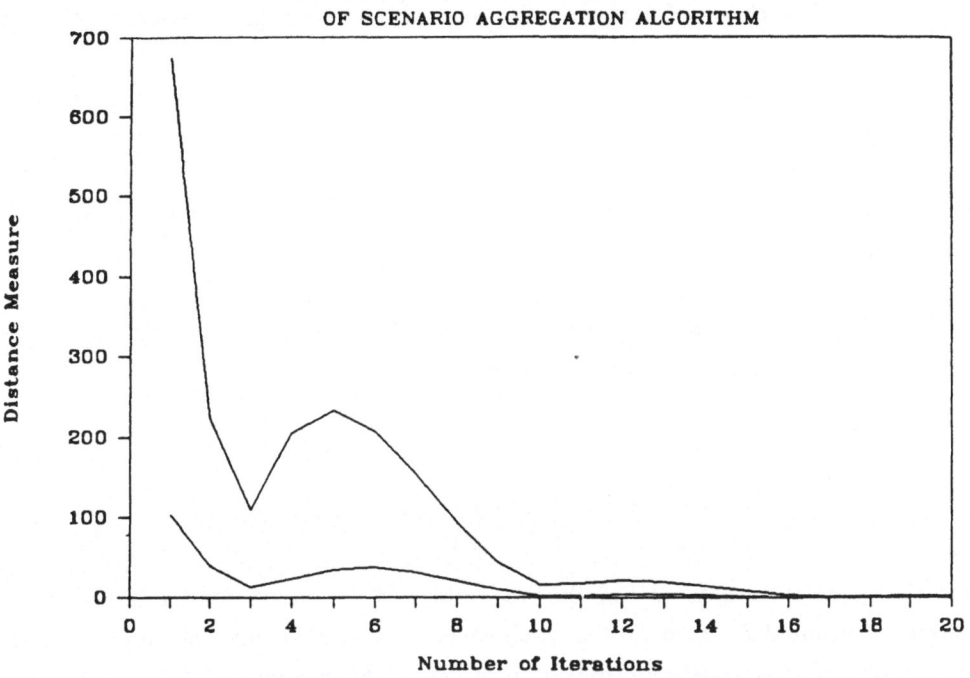

CONVERGENCE PERFORMANCE

OF SCENARIO AGGREGATION ALGORITHM

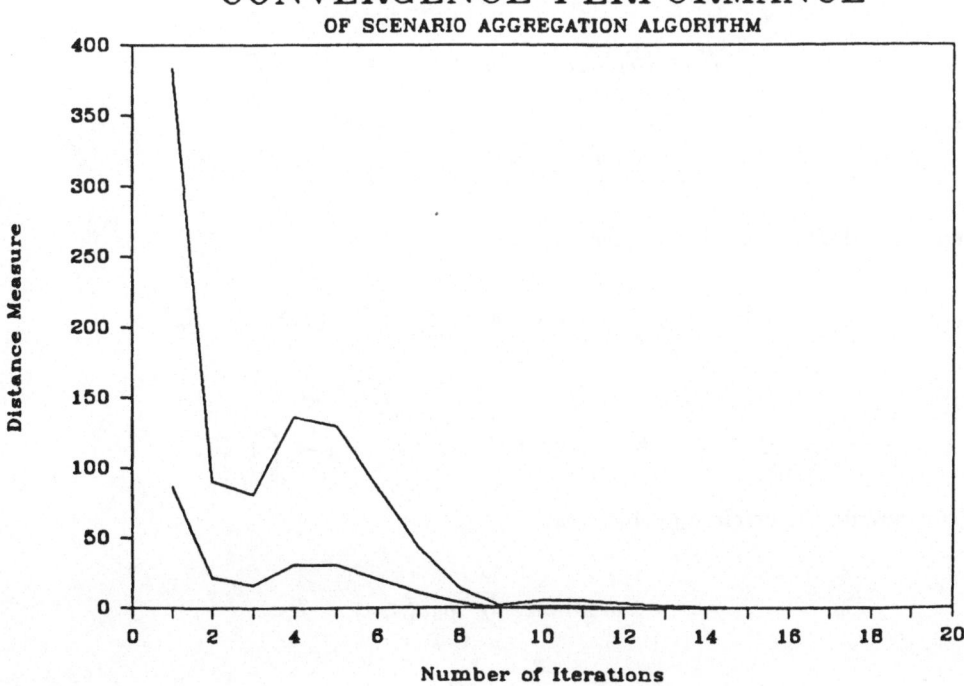

There is still much to do to understand and improve the convergence properties of this algorithm. The choice of the parameter ρ could have a significant effect on the actual performance of the algorithm. The implementation of Mulvey and Hercules [4] seems to confirm this fact. Moreover, much can probably be gained from not carrying out the optimization of the scenario problems to absolute optimality, and restarts could also be used intelligently to minimize the effort required to solve the scenario-problems. When a large number of scenarios is involved, there may also be significant gains to be made from bunching procedures and/or parametric analysis.

5. A Vertical Decomposition for Stochastic Programs

Let us consider the following stochastic programming problem:

$$\text{minimize } E\{f(x,\boldsymbol{\xi})\} = \int f(x,s)\mu(ds)$$
$$\text{subject to } x \in C = \bigcap_{s \in S} C_s,$$

where the space S consists now of the realizations of the random variable $\boldsymbol{\xi}$ and μ is a pobability measure defined on S. This problem is structurally equivalent to problem (1.4). However, for (real) stochastic optimization problems the cardinality of S (the number of possible realizations) will be typically quite large. Let

$$\mathcal{S} := \{S_1, S_2, \ldots, S_N\} \text{ be a partition of } S,$$

and

$$p_n := \int_{S_n} \mu(ds), \quad \text{for } n = 1, \ldots, N.$$

Then, with Bayes' formula we have

$$E\{f(x,\xi)\} = \int f(x,s)\mu(ds) = \sum_n p_n E\{f(x,\xi) \,|\, \xi \in S_n\}.$$

Setting

$$g(x,n) = E\{f(x,\xi) \,|\, \xi \in S_n\} \text{ if } x \in C_n = \bigcap_{s \in S_n} C_s,$$

we can rewrite the original problem as:

$$\text{minimize } \sum_{n=1}^{N} p_n g(x,n).$$

If the number of elements in the partition S is not too large, then this new problem has all the characteristics of the problem for which we suggested a solution method in Sections 2 and 4. Except now each "scenario-problem" is a stochastic programming problem, albeit with a reduced number of possible realizations. In order to make these sub-stochastic programs as easy as possible to solve, it would be to our advantage to choose the partition S so that the realizations that belong to each set S_n be as close as possible to each other. (One may then be able to solve the sub-problems by solving them for just one realization, and proceed by inspection or simple parametric analysis.)

The overall solution scheme would take the following form:

Step 0. Initialize with $\nu = 0$, $w^0(n) = 0$, choose ρ, and $\hat{x}^0$.

Step 1. For $n = 1, \ldots, N$, solve the sub-stochastic program:

$$\text{minimize} \int_{S_n} f^\nu(x, s)\mu_n(ds) \to x_n^{\nu+1}$$

where μ_n is the conditional probability measure on S_n.

Step 2. Update

$$w^{\nu+1}(n) := w^\nu(n) + \rho[x^{\nu+1}(n) - \hat{x}^{\nu+1}]$$

where the "averaged" solution $\hat{x}^{\nu+1}$ is calculated as earlier:

$$\hat{x}^{\nu+1} = \sum_n p_n x^{\nu+1}(n),$$

and return to Step 1 with $\nu = \nu + 1$,

$$f^\nu(x, s) := f(x, s) + w^\nu(n) \cdot x + \tfrac{1}{2}\rho[x - \hat{x}^\nu]^2 \text{ for } s \in S_n.$$

There is, of course, a multiperiod version of this "vertical" decomposition of this algorithm, that involves the same modifications as those introduced in Section 4.

6. Perturbation of the Weights

We have assumed that the weights $\{p_s, \ s \in S\}$ have been fixed in some fashion or another. In the typical scenario-analysis setting, that means that some "expert" has been called in to determine their values. In most situations it will be very useful to study the sensitivity of the solutions to changes in the choice of these weights. From the theory of stochastic programming, see in particular Robinson and Wets [5], we already know that usually the solution of stochastic programs is not very sensitive to minor changes in the probability measure. That theory applies equally well here, and so we do not expect to see major changes in the solutions resulting from minor changes in the values assigned to the weights.

What I shall show here is a simple procedure for "restart", i.e., what to do when a solution x^* (to (1.4)) is available for a particular choice of weights $\{p_s, \ s \in S\}$ and we like to find the solution for the modified weights $\{p'_s, \ s \in S\}$. Let w^* denote the multipliers that are associated with the implementability constraint. The optimality criteria imply:

(i) $x^* \in \operatorname{argmin}[f^*(x,s) \, | \, x \in C_s]$, for all s in S,

(ii) $f^*(x,s) =: f(x,s) + w^* {\cdot} x + \frac{1}{2}\rho\|x - x^*\|^2$

(iii) $\sum_{s \in S} p_s w^*(s) = 0$.

From these it is easy to see that for all $p' = (p'_s, \ s \in S)$ that lie in

$$\mathcal{P}(w^*) := \left\{ p = (p_s, \ s \in S) \, | \, \text{(iii) holds and } \sum_s p_s = 1, \ p_s \geq 0 \right\},$$

the same pair (x^*, w^*) will satisfy the optimality criteria, and the solution will not be affected by this change of weights. The set $\mathcal{P}(w^*)$ is a subset of the $|S|$-dimensional unit simplex, where $|S|$ is the number of elements in S (its cardinality). Since it is determined by $n+1$ equations, recall that n is the dimension of x (and w), it always is a set of dimension $\leq |S| - (n + 1)$. This gives us some latitude in the changes that we can introduce in the weights without having to recompute the solution.

However, when the number of scenarios is small, the set $\mathcal{P}(w^*)$ is generally very low dimensionally (quite often of dimension zero). We then need to recompute a new solution, but as we now indicate, the iterative procedure can easily be started from the solution obtained so far. Indeed let

o $\hat{x}' = x^*$,

o pick some s' such that $0 < p'_{s'} \neq p_{s'}$,

- for all $s \neq s'$, set $w'(s) = w^*(s)$,

- set $w'(s') := (-p'_{s'})^{-1} \sum_{s \neq s'} p'_s w'(s)$.

We still have

$$\sum_{s \in S} p_s w'(s) = 0.$$

Moreover, for all s, except for s',

$$x^* \in \operatorname{argmin}[f'(x,s) \mid x \in C_s], \text{ for all } s \text{ in } S,$$

where

$$f'(x,s) := f(x,s) + w'(s){\cdot}x + \tfrac{1}{2}\rho\|x - x'\|^2.$$

Thus, the first iterate can be found with a minimal amount of effort.

References

[1] Back, Kerry and Stanley Pliska, "The shadow price of information in continuous time decision problems", *Stochastics* **22**(1987), 151–168.

[2] Dempster, Michael, "On stochastic programming: II, Dynamic problems under risk", Research Report DAL TR 86-5, Dalhousie University, July 1985 (revised June 1986).

[3] Evstigneev, Igor V., "Prices on information and stochastic insurance models", IIASA CP-85-23 (Laxenburg), April 1985.

[4] Mulvey, John and Hercules Vladimirou, Solving multistage investment problems: an application of scenario aggregation, Statistics and Operations Research SOR 88-1, Princeton University, 1988.

Mulvey, John and Hercules Vladimirou, Solving multistage stochastic networks: an application of scenario aggregation, Statistics and Operations Report SOR 88-2, Princeton University, 1988.

[5] Robinson, Stephen M. and Roger J-B Wets, "Stability in two-stage stochastic programming", *SIAM J. on Control and Optimization* **25**(1987), 1409–1416.

[6] Rockafellar, R.T. and Roger J-B Wets, "Stochastic Convex Programming: Kuhn-Tucker Conditions", *J. Mathematical Economics* **2**(1975), 349–370.

[7] Rockafellar, R.T. and Roger J-B Wets, "Nonanticipativity and $\mathcal{L}^1$-Martingales in Stochastic Optimization Problems", *Mathematical Programming Study* **6**(1976), 170–187. Also in: Stochastic Systems: Modeling, Identification and Optimization, R. Wets (ed.), North-Holland, Amsterdam, 1976.

[8] Rockafellar, R.T. and Roger J-B Wets, "Scenario and policy aggregation in optimization under uncertainty", IIASA Working Paper WP-87-119, Laxenburg, Austria, December 1987.

[9] Wets, Roger J-B, Large-scale linear programming techniques in stochastic programming, in Numerical Methods for Stochastic Optimization, eds. Y. Ermoliev and R. Wets, Springer Verlag, Berlin, 1987.

METHODS FOR
LARGE-SCALE LINEAR PROGRAMMING*

Walter Murray

Systems Optimization Laboratory, Department of Operations Research
Stanford University, Stanford, California 94305-4022, USA

1 Introduction

The problem we shall address is how to solve the linear programming problem, as stated in the following *standard form*:

$$\text{LP} \qquad \underset{x \in \Re^n}{\text{minimize}} \quad F(x) = c^T x$$
$$\text{subject to} \quad Ax = b,$$
$$x \geq 0,$$

where A is an $m \times n$ matrix ($m < n$). The equation $Ax = b$ describes the *general constraints*, and the inequalities $x \geq 0$ are called *bound constraints*. The essential aim is to find which of the bounds are *binding* at a solution—i.e., which variables x_j are exactly zero.

1.1 Current and new methods

Had this paper been presented four years ago it would have been devoted almost entirely to a description of the simplex method (Dantzig [Dan63]). The advent of Karmarkar's projective method [Kar84] has resulted in an enormous "resurgence" of interest in alternative methods to solving LP. We say resurgence because during the early years after the discovery of the simplex method a wide variety of methods were proposed. Indeed methods similar to Karmarkar's were among them (e.g. Motzkin [Mot51]). Given the computing equipment and the sparse-matrix technology of the time, all of these early methods slipped into disuse. It is premature to assume that the new methods will entirely replace the simplex method or even become the dominant approach. What has become clear is that, at least for some problems, a competitive alternative to the simplex method is now available. One virtue of the new approaches is that they are radically different from the simplex method. It therefore seems likely that they will be complementary to the simplex method. It may be that a hybrid approach combining the new methods with the simplex method will also prove effective.

   In his original paper, Karmarkar was concerned with proving that the projective method had a computational complexity that was polynomial in n. (In particular, he showed that the

---

*The material contained in this paper is based upon research supported by the Air Force Office of Scientific Research Grant 87-01962; the U.S. Department of Energy Grant DE-FG03-87ER25030; National Science Foundation Grants CCR-8413211 and ECS-8715153; and the Office of Naval Research Contract N00014-87-K-0142.

NATO ASI Series, Vol. F51
Algorithms and Model Formulations
in Mathematical Programming
Edited by Stein W. Wallace
© Springer-Verlag Berlin Heidelberg 1989

bound on computation time was of order $n^{3.5}L$, where L is a measure of the required accuracy in the solution.) While this is of considerable theoretical importance, it would be wrong to attribute the success of the new methods to this property. The polynomial bound obtained (and subsequently improved upon) is for all practical purposes still astronomically large. Indeed the most commonly implemented method the so-called affine variant (see [VMF8]) is thought not to be polynomial. The fact that the methods perform much better on practical problems than their theoretical limitations is what makes the methods interesting. This is precisely the case with the simplex method.

The current limitation of the new methods is the need to solve a large, sparse least-squares problem (or its equivalent). It is not difficult to construct examples where the simplex basis is easy to factorize, yet there is no known way of solving the corresponding least-squares problem with similar efficiency. Even in their method of failing, the new approaches differ from simplex. If it is possible to perform one iteration of the new methods it is likely that the problem can be solved—in other words, the potential difficulty is not being able to perform a single iteration. Conversely, with the simplex method it is almost always possible to perform an iteration, but the number of iterations required to determine the solution may be too large. One avenue of research being explored in the new approaches is how to reduce the amount of work required to perform an iteration without unduly increasing the number of iterations.

1.2 Dense and sparse problems

Linear programs may be divided into three categories: dense problems, sparse problems, and structured problems. A problem is said to be *sparse* if there is a significant proportion of zero elements in A. It is tacitly assumed that if a problem is large then it is sparse. Structured problems occur when explicit advantage may be taken of the *pattern* of the nonzero elements. When the structure of the nonzeros is sufficiently simple (as with network problems) it is possible to develop very efficient forms of the simplex algorithm. We shall not be concerned with such algorithms here. It should be noted that all practical sparse problems have *some* pattern to the occurrence of nonzero elements. If that were not the case (i.e., if the pattern were random), sparse-matrix techniques would be of little use.

When there is a choice, a question often asked is whether it is better to solve a large sparse problem or a small dense one. Suppose the number of nonzeros in the sparse problem is M and the size of A in the dense problem is $m \times n$. A reasonable choice is to solve the dense problem when $mn < 5M$. The bound is not critical since the two approaches will be similar in efficiency if $mn \approx 5M$. This bound assumes that a conventional computer is to be used. For vector machines or machines with many parallel processors it may be worthwhile to solve the dense problem for much larger factors of M. The advent of new architectures has renewed interest in solving problems that are assumed to be dense. In general, problems become progressively more sparse with increasing size, and for large examples the density of nonzero elements is usually less then 1%. The average number of nonzeros in a column is typically less than 10, regardless of the size of the problem.

1.3 Summary

We shall assume the reader has a basic knowledge of linear programming and is familiar with such terms as feasible points, Lagrange multipliers, etc.

Sections 2–4 are devoted to the simplex method. They describe the method from an "active-set" point of view, and discuss various implementation aspects.

Sections 5–7 describe the use of a logarithmic barrier function for solving problem LP. The approach is simply a direct application of Newton's method for linearly constrained optimization. The resulting algorithm is closely related to Karmarkar's projective method.

Some related methods are briefly reviewed in the remaining sections.

2 The Simplex Method

We shall assume for the moment that all the vertices of the feasible region are non-degenerate. (The consequences of degeneracy are discussed at the end of this section and in Section 4.2.) Under such circumstances it can be shown that if an optimal solution exists for LP, there is a solution at a vertex. Hence, if it were known which constraints formed the vertex, the solution could be found by solving one set of linear equations. The simplex method may be viewed as a procedure for systematically estimating the set of constraints that are binding at the solution. More specifically, if the current estimate is a feasible (but non-optimal) vertex, the next estimate is chosen as an adjacent feasible vertex at which the objective function has a lower value. An iteration of the simplex method consists of solving two systems of linear equations (one to determine the choice of vertex and one that computes the vertex) that differ only slightly from those solved at the previous iteration.

We term the estimate of the set binding constraints at the solution the *working set*. Let A_W denote the matrix of constraints in the working set. In the case of LP in standard form, the working set includes the general equality constraints $Ax = b$ plus a subset of the bounds. We assume that A_W is square and nonsingular and x, the current estimate of an optimal vertex x^*, is feasible.

For simplicity, we assume the variables have been rearranged so that the last $n - m$ bounds are exactly satisfied at the current iterate x, i.e., the last $n - m$ components of x are zero. Then A and x may be partitioned as

$$A = \begin{pmatrix} B & N \end{pmatrix}, \qquad x = \begin{pmatrix} x_B \\ x_N \end{pmatrix}, \tag{1}$$

where B is an $m \times m$ matrix. We shall assume an initial working set is known such that B is *nonsingular* and x is feasible, so that $x_B > 0$ and $x_N = 0$. The matrix B in (1) is called the *basis*. The variables x_B associated with B are called the *basic* variables, and the variables x_N are the *nonbasic* variables. When A has the form (1), the working set (denoted by A_W) is given by

$$A_W = \begin{pmatrix} B & N \\ 0 & I \end{pmatrix}. \tag{2}$$

Since B is nonsingular it follows that A_W is nonsingular. From the definition of the working set and (2), x must satisfy

$$A_W x = \begin{pmatrix} B & N \\ 0 & I \end{pmatrix} \begin{pmatrix} x_B \\ x_N \end{pmatrix} = \begin{pmatrix} b \\ 0 \end{pmatrix}. \tag{3}$$

To determine whether x is optimal for the original linear program, we evaluate the Lagrange multipliers λ, which satisfy

$$A_W^T \lambda = c. \tag{4}$$

The vector λ is again partitioned into π, the m-vector of multipliers associated with the equality constraints $Ax = b$, and η, the $(n - m)$-vector of multipliers associated with the bound constraints. Using (2), equation (4) may then be written as

$$c = \begin{pmatrix} c_B \\ c_N \end{pmatrix} = \begin{pmatrix} B^T & 0 \\ N^T & I \end{pmatrix} \begin{pmatrix} \pi \\ \eta \end{pmatrix} = \begin{pmatrix} B^T \pi \\ N^T \pi + \eta \end{pmatrix}, \tag{5}$$

where c_B denotes the elements of c corresponding to the basic variables, and c_N denotes the elements of c corresponding to the nonbasic variables. It is immediate from (5) that π solves the nonsingular linear system

$$B^T \pi = c_B. \tag{6}$$

The multipliers associated with the bounds may then be computed from the remaining equations of (5), which give

$$\eta = c_N - N^T \pi. \tag{7}$$

In LP terminology, the Lagrange multipliers η are known as the *reduced costs*, and the process of computing them is known as *pricing*.

If $\eta \geq 0$ then x satisfies the optimality conditions for LP and the optimal vertex has been found. Otherwise, there exists an i for which $\eta_i < 0$. Since A_W is nonsingular a vector p exists such that

$$A_W p = e_{m+i}.$$

By definition it is a feasible direction for the constraints in the working set. Let

$$\bar{x} = x + \alpha p,$$

where α is the largest feasible step along p. Since $c^T p = c^T A_W^{-1} e_{m+i} = \eta_i < 0$ it follows $c^T \bar{x} < c^T x$. Let the p be partitioned as p_B and p_N. We have

$$A_W p = \begin{pmatrix} B & N \\ 0 & I \end{pmatrix} \begin{pmatrix} p_B \\ p_N \end{pmatrix} = e_{m+i}. \tag{8}$$

Since $p_N = e_i$, rearrangement of (8) shows that p_B satisfies

$$B p_B = -N p_N = -a_{m+i}, \tag{9}$$

where a_{m+i} is the *column* of A corresponding to the new nonbasic variable.

Given the search direction p defined by (8), the step α to the nearest constraint is the distance along p at which one of the basic variables reaches its lower bound of zero. The step α is given by

$$\alpha = \min\{-x_j/p_j \mid p_j < 0\}.$$

At $\bar{x} = x + \alpha p$, a new bound constraint is added to the working set, with the result that *different* columns of A will compose B and N at the next iteration. Each iteration of the simplex method has the effect of "swapping" a basic and a nonbasic variable, which is why LP terminology refers to a change in the working set as a "change of basis". When applied to the standard form LP, the simplex method requires changes to only the columns (not the rows and columns) of B. It is for this reason that conversion to the standard form is worthwhile for large problems.

We need to show that the conditions at the start of the next iteration correspond to those at the first. Previously we assumed that A_W was nonsingular and x was feasible. Clearly the new estimate of the solution is feasible. It remains to be shown that $\bar{A}_W$, the matrix for the new working set, is nonsingular. Let $\hat{A}$ denote the $(n-1) \times n$ matrix composed of the rows of $\bar{A}_W$ in A_W. Consider the following lemma.

Lemma 2.1. *Let V be an $(n-1) \times n$ matrix of rank $n-1$. Let a be a vector such that $a^T p \neq 0$. If $Vp = 0$ then $\bar{V}^T \equiv \left(\begin{array}{cc} V^T & a \end{array} \right)$ is nonsingular.*

Proof. Since V has rank $n-1$ it follows that $\bar{V}$ can be singular only if there exists u such that $a = V^T u$. Premultipling by p^T gives

$$p^T a = p^T V^T u = 0.$$

This contradicts the assumption that $a^T p \neq 0$. It follows that $\bar{V}$ is nonsingular. ∎

Although in the case of the standard form LP the working set has a special form, the nonsingularity of the matrix of the working set does not depend on the special form. It is easily seen from (8) that A with the $(m+i)$th row removed satisfies the conditions on V. Let e_j^T be the new row of $\bar{A}_W$. Since p intercepts the new constraint in the working set we must have $p_j = e_j^T p \neq 0$. It follows from Lemma 2.1 that $\bar{A}_W$ is nonsingular.

2.1 A summary of the steps of the simplex algorithm

Choose x to be a feasible vertex with associated nonsingular basis B.

1. Compute the reduced costs by solving

$$B^T \pi = c_B$$

   and setting $\eta = c_N - N^T \pi$.

2. Identify an index i such that $\eta_i < 0$. The associated nonbasic variable will become basic. If none exists then terminate—the current x is optimal.

3. Set $p_N = e_i$ and solve

$$Bp_B = -a_{m+i}.$$

4. Compute the maximum feasible step along p:

$$\alpha = \min\{-x_j/p_j \mid p_j < 0\}.$$

   Identify the associated basic variable. This variable will become nonbasic.

5. Set $x \leftarrow x + \alpha p$.

6. Redefine the set of basic and nonbasic variables, i.e., reorder the variables and the columns of A.

Observe that the choice of index of both the incoming and outgoing basic variable may not be unique. The outgoing basic variable *will* be unique if there are no degenerate vertices. How best to break ties depends on several criteria, some of which are discussed in Section 4.

2.2 Convergence

Convergence of the simplex method will occur in a finite number of iterations provided no degenerate vertex is encountered (assuming that the objective function is bounded below). The objective function is then strictly decreasing, so that no choice of working set can be repeated. Since there is a finite number of possible choices, we must identify the correct working set after a finite number of iterations.

Degenerate vertices pose a problem because the step to the nearest constraint not in the working set may be zero. (The objective function is then not reduced.) At such points there is a danger of *cycling*. Cycling is the term used to describe the occurrence of a repeat basis (one that has occurred earlier). Once a basis is repeated the whole set of intervening bases will repeat ad infinitum.

Many procedures have been designed to avoid cycling. They can be divided into two types: those based on lexicographical ordering, and those based on perturbing the constraints (thereby destroying degeneracy at the cost of creating many more vertices). For cycling to occur we must have $\alpha = 0$ and a multiple choice for the new basic variable. A particularly simple lexicographical procedure is due to Bland [Bla77]. In Bland's scheme both the leaving and entering basic variables are chosen to be the eligible variable of smallest index. Under such a rule it can be shown that no basis will be repeated.

Cycling is not a serious difficulty when solving practical LP problems. However, degeneracy is common and schemes to handle degeneracy efficiently are still a subject of research. A drawback of schemes such as Bland's is that they give equal weight to all variables. The number of possible choices of basic variables at a degenerate vertex may be exceedingly large. We need to be able to choose basic variables that have a better-than-even chance of being in the set that allows a move from the current vertex. In addition we would like to choose a sequence of well-conditioned bases.

3 Finding a Feasible Point

In describing the simplex algorithm we have made the assumption that a feasible vertex is known. Ordinarily this will not be the case and it is necessary to be able to compute one.

The general approach adopted to finding feasible points (they need not be vertices) is to define a nonnegative measure of infeasibility and then minimize it. If a feasible point exists then the measure is reduced to zero. The choice of measure depends on the form the constraints, the type of feasible point required, and the initial information available about the constraints. For example, in the simplex method we require a vertex and a factorization of the initial basis (to enable the equations in step 1 and step 2 to be solved)

For the simplex method a technique known as *Phase 1* is usually the method of choice. The optimality phase (minimizing the objective function while remaining feasible) is often referred to as *Phase 2*. We consider the problem in standard form but with upper and lower bounds:

$$Ax = b, \qquad l \leq x \leq u.$$

When the solution is to be determined by the simplex algorithm, we need a feasible *vertex*. Often an initial basis and its factorization are already known. If not, we can choose a set of basic variables and attempt to factorize the corresponding set of columns. During the factorization it will be revealed whether or not the basis is nonsingular. If singularity does occur (and this is common), certain basic variables can be replaced by judiciously chosen slack variables.

Once a nonsingular basis factorization is known, a point satisfying the general constraints may be computed by setting the nonbasic variables at one of their bounds. The only possible constraints that can be violated are the bounds on the basic variables. A measure of infeasibility is therefore

$$F(x) = \sum_{j \in \mathcal{J}_l} (l_j - x_j) + \sum_{j \in \mathcal{J}_u} (x_j - u_j),$$

where $\mathcal{J}_l$ and $\mathcal{J}_u$ are the sets of indices of the variables that violate their lower and upper bounds respectively. The function $F(x)$ is *linear* and is equal to the *sum of infeasibilities* at x. Note that $F(x)$ is zero at any feasible point, and positive otherwise. Therefore, a feasible point can be found by minimizing $F(x)$ subject to the constraints $Ax = b$ and the bounds on the variables that are currently satisfied (including all the nonbasic variables). This is a linear programming problem (or almost) for which a feasible vertex is known. The set of variables satisfying their bounds may change at each iteration (it can only stay the same or increase). The linear objective function is only piecewise linear and hence the object function may change at each iteration. Fortunately, none of these differences materially affects the working of the simplex algorithm.

The Phase-1 linear program is typically solved by a slightly modified version of the simplex method. We first note that it is necessary to amend the rule of choosing α to be the largest feasible step. (Consider the case when all the lower bounds are violated, all the elements of p are non-negative and there are no upper bounds.) We could choose the step to be the smallest one for which a basic variable hits a bound (such an event must occur). However, if the first basic variable to hit its bound is currently infeasible then a larger step would make it feasible and basic (again).

When the infeasible set changes, there is a discontinuity in the gradient of $F(x)$ and p may cease to be a descent direction. However, this may not occur until several variables have been made feasible along a given direction. Therefore, a second strategy is to choose the step that reduces $F(x)$ as much as possible along p subject to no further bounds being violated. A third strategy is to ignore the change to $F(x)$ and to take the step along p that reduces the maximum *number* of infeasibilities, subject to satisfying the currently satisfied bounds. In the last two strategies the step may not be unique. In both cases the smallest satisfactory step is taken.

It is easy to appreciate that on any given iteration the second strategy, although it reduces $F(x)$ by a greater amount, usually does not reduce the number of infeasibilities by as much as the last strategy. In practice the last strategy is usually preferred. It often determines a feasible point in fewer iterations and the effort to compute the required step is less. However, there are problems where the second strategy is more effective.

It may be seen that at each iteration either $F(x)$ or the number of variables violating their bounds is reduced. Since neither $F(x)$ nor the number of infeasibilities can be reduced indefinitely, the algorithm must terminate after a finite number of iterations. As before it is necessary to avoid cycling at degenerate vertices. If a point is reached such that all the Lagrange multipliers are positive but some bounds are still violated, no feasible point exists. When the sum of infeasibilities is reduced to zero, a feasible vertex and a basis factorization will be available to commence Phase 2.

Although we have described the feasible-point algorithm in terms of the standard-form problem, the same principle can be applied to any form of LP. For example, if we have general inequalities $Ax \geq l$ and some are violated, we include a term in the objective function of the form

$$F(x) = \sum_{j \in \mathcal{J}} l_j - a_j^T x,$$

where $\mathcal{J}$ is the set of violated general constraints. Again this is a piecewise linear function. If a dense problem is being solved and no point is known satisfying the general constraints, the usual approach is to choose an initial point that satisfies the bounds on the variables.

3.1 Minimizing the sum of infeasibilities

The procedure described will not minimize the sum of infeasibility if a feasible point does not exist. Usually when we discuss the minimum of the infeasibilities we mean one of two things. For the standard-form problem LP we mean the minimum violation of the *bounds* subject to the general constraints being satisfied. When the problem is not converted to standard form, we mean the minimum sum of the general inequalities subject to the bounds on the variables (if they exist) being satisfied.

Suppose we apply the procedure described and reach a point where all the multipliers are positive, yet $F(x) > 0$. Let the subset of constraints in the working set be denoted by A_w. It may be possible to reduce $F(x)$ by moving infeasible with respect to one of the constraints in the working set. Although this adds another residual to $F(x)$, the increase in this residual may be more than offset by the reduction in the other residuals.

Suppose we move infeasible with respect to the i-th lower bound, so that

$$A_w p = -e_i.$$

The current objective function is given by

$$F(x) = c^T x + \sum_{j \in \mathcal{J}_l} l_j - x_j + \sum_{j \in \mathcal{J}_u} x_j - u_j,$$

where $i \notin \mathcal{J}_l \cup \mathcal{J}_u$. Since F has a discontinuity in its derivative along p at $\alpha = 0$, we need to show that $\bar{c}^T p < 0$, where $\bar{c} = c - e_i$. From the definition of p we have

$$\bar{c}^T p = -(c - e_i)^T A_w^{-1} e_i = -\eta_i + 1,$$

where η_i is the reduced cost of the i-th bound. It follows that $\bar{c}^T p < 0$ if $\eta_i > 1$.

We may find the minimum sum of infeasibilities by first applying the feasible-point routine previously described. If no feasible point exists we may then choose new basic variables from any nonbasic variable whose reduced cost is larger than one. At this stage we must choose the step in a way that reduces $F(x)$. (Reducing the number of infeasibilities is no longer satisfactory.) It may be that negative reduced costs re-occur; however, it is still necessary to choose a step that reduces $F(x)$. After a finite number of iterations, the minimum sum of infeasibilities will be determined and the reduced costs will satisfy $0 \leq \eta \leq e$.

One context in which it is desirable to minimize the sum of infeasibilities is when the LP occurs as a subproblem in a decomposition algorithm such as that of Benders [Ben62]. Also, it may be that the *minimum* $F(x)$ is small, yet when it is first detected that no feasible point exists, $F(x)$ could be large. Indeed we may decide that an *acceptable* feasible point has been determined. In solving practical problems it is necessary to allow small infeasibilities in the basic variables. If the minimum sum of infeasibilities is found it can happen that none of the bounds are exceed by more than the feasibility tolerance. If this is not the case it may indicate the feasibility tolerance is too small.

4   Computational Aspects of the Simplex Method

The algorithm as described does not define how to resolve ties in the choice of both incoming and outgoing basic variables. In order to obtain rapid convergence it is necessary to make a "good" choice for the incoming basic variable (the eligible set is often very large). Even with a good choice it is possible to generate problems for which every vertex is visited before the optimal vertex is determined. Despite such examples, the simplex method with a good choice of initial basic variables normally converges in a small multiple of m iterations (typically 2 to 4) even from a poor initial estimate. In practice, a good starting point is often known (the solution of a similar problem) and the number of iterations is even smaller.

4.1   Choosing the new basic variable

For dense problems the usual choice is to pick the variable with the smallest reduced cost. The reasoning behind this strategy is that such a choice will reduce the objective function by the greatest amount for a fixed change in the basic variable chosen. The only other choice that has received serious consideration is to maximize $\beta_i = -c^T p^i / \|p^i\|$, where p^i is the search direction corresponding to making the ith nonbasic variable basic. This is referred to as the *steepest-edge strategy*. One virtue of the strategy is its invariance to scaling. In most tests the steepest-edge strategy finds a solution in fewer iterations. Unfortunately we need to compute β_i for all indices for which $\eta_i < 0$. This does not mean computing p^i for all choices of i (as it may seem at first sight). However, the work per iteration does increase, perhaps as much as 50%. The current consensus, though not unanimous, is that the saving in terms of reduced iterations (which may on occasion not happen) usually does not outweigh the extra work per iteration (which is bound to occur). It seems unlikely that more elaborate schemes would be worthwhile, such as finding the adjacent vertex with the lowest objective function.

Interest in steepest-edge strategies has been revived recently by Forrest and Tomlin [FT88]. It turns out that on vector machines the relative effort for such schemes is much smaller.

It can be observed from the definition of η that individual elements may be computed once π is known. For large problems it is sometimes worthwhile computing only a few elements of η and checking whether a suitable (sufficiently negative) element has been determined. Such procedures are known as *partial pricing*. The savings are particular significant if n is large compared to m (say $n \geq 5m$). Although the motivation for such procedures is to reduce the work per iteration (with the hope that the number of iterations will not increase significantly), the observed result is that the number of iterations is often reduced significantly. It not clear why this should be so. One possible explanation is that adjacent columns of A are often correlated (in terms of scale, role, etc.). Partial pricing forces a selection both within and outside any particular set of columns. Partial pricing may be implemented in a number of ways. One possibility is to divide the columns into groups and in successive iterations consider a choice from successive groups. If the number of groups is k then after at most k iterations all nonbasic variables have been considered.

4.2   Choosing the nonbasic variable

Unless the current or following vertex is degenerate there is no choice of outgoing basic variable. However, a degree of choice can be deliberately introduced to try and ensure that B is not ill-conditioned. Suppose instead of the step to the precise boundary of the feasible region we compute the step to a point that violates any given constraint by some small quantity, say δ. Let this step be denoted by $\bar{\alpha}_i$. We have

$$\bar{\alpha} = \min\{\bar{\alpha}_i = -(x_i + \delta)/p_i \mid p_i < 0\}.$$

We then choose the outgoing variable to be one that reaches its bound at a step no larger than $\bar{\alpha}$, with largest possible "pivot element" p_i:

$$p_r = \max\{ |p_i| \mid p_i < 0 \text{ and } \alpha_i \le \bar{\alpha}\},$$

where $\alpha_i = -x_i/p_i$ is the step to the (zero) lower bound on x_i. In general, the index r is different from the one that defines $\bar{\alpha}$ or the original α. The step ultimately taken is $\alpha = \max\{\alpha_r, 0\}$.

If we choose the outgoing variable according to this rule, we cannot violate the other basic variables by more than δ. To appreciate why the resulting basis is usually better conditioned, let $\bar{B}$ be the new basis. It follows from (9) that a vector v exists such that

$$\bar{B}v = p_r a_r,$$

where r is the outgoing basic variable. Obviously if p_r is close to zero, $\bar{B}$ may be nearly singular. We may not expect p_r to be close to zero, since this suggests that the resulting step is large. However, it may be that x_r is also close to zero. Consider the following example.

Let $x_1 = -10^{-10}$, $x_2 = 1.2$, $p_1 = -10^{-10}$, and $p_2 = -1$. Using the original rule we would choose x_1 as the incoming variable and $\alpha = 1$. Using the second rule with $\delta = 10^{-8}$ we would choose the second variable with a step of 1.2. If we made the first choice, the new basis could be almost singular. With the second choice this danger does not exist. Although the bound on the first variable is violated the degree of violation is less than 10^{-8}. The error in x and η depends on the conditioning of B. By introducing a controlled infeasibility in x (usually referred to as a feasibility tolerance) the computational error in x and η may be considerably reduced.

The scheme just described is due to Harris [Har73]. More recently, Gill *et al.* [GMSW88] have described an alternative procedure that allows a *variable* feasibility tolerance. As the iterations proceed, the degree of infeasibility is allowed to increase (albeit very slowly). Such a strategy always permits the objective function to be decreased and a positive step to be taken. Consequently the increase in condition number of the basis is strictly bounded. None of these properties holds for the Harris scheme.

Although all the properties are worthwhile, the main motive for introducing a variable feasibility tolerance is to improve the efficiency of the simplex method at degenerate vertices. The scheme described in [GMSW88] reduces the likelihood of cycling to a negligible probability.

4.3 Computing the basis factorization

The main computational effort in the simplex method is to solve the system of equations arising in steps 1 and 3. Systems of equations involving B and B^T are typically solved using an LU factorization of B, obtained by Gaussian elimination. That is, matrices L and U are found such that $B = LU$, where L is lower triangular and U is upper triangular. The initial basis must be factorized directly, but for subsequent bases the factors are usually updated from the current factorization. Periodically there is a need to discard the updated factors and refactorize.

When B is sparse, sparse LU factors may be found by a judicious permutation of the rows and columns of B. There is no known practical way to determine a permutation that literally minimizes the number of nonzero elements in L and U. One common approach is to find a permutation of B such that the permuted form is almost lower triangular. A second approach is to choose the pivot during each stage of Gaussian elimination according to some measure of merit. For example, the product of nonzero elements in the pivot row and pivot column is often minimized (excluding the pivot row and column themselves); i.e., $\min(r_i - 1)(c_j - 1)$ among eligible rows i and columns j (a popular rule due to Markowitz). Fill-in can only occur in those rows for which there is a nonzero element in the pivot column. Moreover, the degree of fill-in in each row being modified is less than the number of nonzeros in the pivot row. It can be appreciated, therefore, that this measure of merit helps to reduce fill-in.

In all schemes it may be necessary to override the ordering dictated by the wish to reduce fill-in, in order to ensure numerical stability. Which approach works best depends on the nature of the problem. It can be shown that sparse matrices whose nonzero elements occur at random locations have almost dense LU factors irrespective of the permutation of the rows and columns used. Experience with practical LP problems has shown that the basis may be factorized with very little fill-in. Indeed it is usual for the total number of nonzero elements in L and U to be less than double the number in B.

Given an initial factorization, it is relatively efficient to obtain a factorization of the next basis. There are several ways carry out the necessary updating. In almost no case is an *explicit* lower triangular factor found. (An exception is the method of Fletcher and Matthews [FM84], which has not been implemented in the sparse case.) A description of the various updating techniques commonly employed may be found in Gill *et al.* [GMSW84]. All updating techniques add to the original data file required to store the LU factors. It is therefore necessary to discard the factors when the growth of the file storing the factors threatens to exceed certain limits, or the time to perform an iteration (which increases when the data file grows) makes refactorization worthwhile.

On refactorization it is worthwhile recomputing the current estimate of the solution, to ensure that $Ax = b$ is satisfied. It can happen that the recomputed basic variables no longer satisfy their upper or lower bounds, as a result of numerical error in the preceding iterations. If so, it is necessary to reinstitute the feasibility procedure.

5 A Barrier-Function Approach

There are many ways of viewing the new methods that have been motivated by Karmarkar's projective algorithm (to some degree that is part of their fascination). Karmarkar's own approach was to give a geometric interpretation. We have chosen to interpret the methods in terms of *barrier functions*. Almost all the new approaches have such an interpretation, hence barrier functions give a unity to methods that at first sight may seem quite disparate. The differences between algorithms can usually be shown to result from different choices of the *barrier parameter* (see 10).

Barrier-function methods treat *inequality* constraints by creating a barrier function, which is a combination of the original objective function and a weighted sum of functions with a positive singularity at the constraint boundary. (Many barrier functions have been proposed; we mainly consider the logarithmic barrier function, first suggested by Frisch [Fri55].) As the weight assigned to the singularities approaches zero, the minimum of the barrier function approaches the minimum of the original constrained problem. Barrier-function methods require a strictly feasible starting point for each minimization, and generate a sequence of strictly feasible iterates.

(For a complete discussion of barrier methods, see Fiacco [Fia79]; both barrier and penalty function methods are described in Fiacco and McCormick [FM68]. Brief overviews are given in Fletcher [Fle81] and Gill, Murray and Wright [GMW81].)

The original idea of barrier and penalty functions was to eliminate *nonlinear* constraints. At the time they were proposed, efficient algorithms for nonlinearly constrained problems were not known. A number of authors, notably Fiacco and McCormick, also investigated their use in solving LP's. Such suggestions never took root. The rationale for the use of barrier functions in LP is different from that when applying them to nonlinear problems. The hope in the LP case is to avoid the inherent combinatorial aspect of the simplex algorithm.

In the original barrier algorithms, the approach was to select a value for the barrier parameter (the weight placed on the barrier term) and minimize the resulting function. The barrier parameter was then reduced and the process was repeated. A modification to this approach adopted in [Mur69,Wri76,MW78] was to vary the parameter at every iteration. Such algorithms attempted to follow the *trajectory* of minimizers. They generated a sequence of iterates whose distance to the trajectory was proportional to the distance to the solution. It will be seen that the trajectory approach is perhaps the one most suitable for the LP case.

5.1 Applying a barrier transformation to a linear program

When a logarithmic barrier-function method is applied to LP, the subproblem to be solved is:

$$\underset{x \in \Re^n}{\text{minimize}} \quad F(x) \equiv c^T x - \mu \sum_{j=1}^{n} \ln x_j \tag{10}$$
$$\text{subject to} \quad Ax = b,$$

where the scalar μ ($\mu > 0$) is known as the *barrier parameter* and is specified for each subproblem. The equality constraints $Ax = b$ cannot be treated by a barrier transformation, and thus are handled directly.

If $x^*(\mu)$ is the solution of (10), then $x^*(\mu) \to x^*$ as $\mu \to 0$ (see, e.g., Fiacco and McCormick [FM68]). Very strong order relations can be derived concerning $x^*(\mu)$ and $c^T x^*(\mu)$ (see, e.g., Mifflin [Mif72,Mif75], Jittorntrum [Jit78], Jittorntrum and Osborne [JO78]). In particular, when LP is primal nondegenerate,

$$\|x^*(\mu) - x^*\| = O(\mu) \tag{11}$$

for sufficiently small μ. When LP is dual degenerate, the corresponding relation is

$$\|x^*(\mu) - x^*\| = O(\sqrt{\mu}).$$

Under these circumstances, Jittorntrum and Osborne [JO80] have suggested a modification to the barrier function (they introduce weights on the log terms) that retains the relationship (11).

The optimality conditions for (10) imply that at $x = x^*(\mu)$, there exists a vector $\pi(\mu)$ such that

$$c = A^T \pi(\mu) + \mu D^{-1} e, \tag{12}$$

where

$$D = \text{diag}(x_j), \quad j = 1, \dots, n, \tag{13}$$

and $e = (1, 1, \dots, 1)^T$. If LP is primal nondegenerate, $\pi(\mu) \to \pi^*$ as $\mu \to 0$, and

$$\lim_{\mu \to 0} \frac{\mu}{x_j^*(\mu)} = \eta_j^*. \tag{14}$$

The traditional approach to barrier methods is to minimize the barrier function for a sequence of decreasing positive values for μ, the final value of μ being sufficiently close to zero to make the minimizer of the barrier function a satisfactory solution of the original problem. It is unnecessary to find the minimizer exactly for the intermediate values of μ. Therefore, an optional strategy is to perform only a limited number of steps of the minimization (perhaps only one) before reducing μ.

5.2   Solution of the subproblem

Given a linearly constrained problem of the form

$$\underset{x}{\text{minimize}} \ F(x) \quad \text{subject to} \quad Ax = b,$$

a standard approach is to use a *feasible-point descent method* (see, e.g., Gill, Murray and Wright [GMW81]). The current iterate x always satisfies $Ax = b$, and the next iterate $\bar{x}$ is defined as

$$\bar{x} = x + \alpha p, \tag{15}$$

where p is an n-vector (the *search direction*) and α is a positive scalar (the *steplength*). The computation of p and α must ensure that $A\bar{x} = b$ and $F(\bar{x}) < F(x)$.

The *Newton search direction* associated with (10) is defined as the step to the minimum of the quadratic approximation to $F(x)$ derived from the local Taylor series, subject to retaining feasibility. Thus, the Newton search direction p is the solution of the following quadratic program:

$$
\begin{aligned}
\underset{p \in \Re^n}{\text{minimize}} & \quad g^T p + \tfrac{1}{2} p^T H p \\
\text{subject to} & \quad A p = 0,
\end{aligned}
\tag{16}
$$

where $g \equiv \nabla F(x)$ and $H \equiv \nabla^2 F(x)$. If π is the vector of Lagrange multipliers for the constraints in (16), then the required solution satisfies the linear system

$$
\begin{pmatrix} H & A^T \\ A & 0 \end{pmatrix} \begin{pmatrix} -p \\ \pi \end{pmatrix} = \begin{pmatrix} g \\ 0 \end{pmatrix}. \tag{17}
$$

Note that π converges to the Lagrange multipliers for the constraints $Ax = b$ in the original problem LP.

When $F(x)$ is the barrier function in (10), its derivatives are

$$g(x) = c - \mu D^{-1} e \quad \text{and} \quad H(x) = \mu D^{-2},$$

where D is defined by (13). Note that g and H are well defined only if $x_j \neq 0$ for all j. Since $H(x)$ is positive definite when $x > 0$, p is finite and unique, and is a *descent direction* for $F(x)$, i.e., $(c - \mu D^{-1} e)^T p < 0$. It also implies that $F(x)$ is a strictly convex function for strictly feasible x and that $F(x)$ has a unique minimizer.

It follows from (17) that p and π satisfy the equation

$$
\begin{pmatrix} \mu D^{-2} & A^T \\ A & 0 \end{pmatrix} \begin{pmatrix} -p \\ \pi \end{pmatrix} = \begin{pmatrix} c - \mu D^{-1} e \\ 0 \end{pmatrix}. \tag{18}
$$

Rewriting (18) in terms of a vector r defined by $Dr = -\mu p$, we see that r and π satisfy

$$\begin{pmatrix} I & DA^T \\ AD & 0 \end{pmatrix} \begin{pmatrix} r \\ \pi \end{pmatrix} = \begin{pmatrix} Dc - \mu e \\ 0 \end{pmatrix}. \tag{19}$$

It follows that π is the solution and r the optimal residual of the following linear least-squares problem:

$$\underset{\pi}{\text{minimize}} \ \|Dc - \mu e - DA^T\pi\|_2. \tag{20}$$

The projected Newton barrier direction is then

$$p = -(1/\mu)Dr. \tag{21}$$

For a given positive μ, Newton's method will eventually reach a domain in which the "ideal" unit step along the direction p will be feasible and reduce the barrier function. The iterates can thereafter be expected to converge quadratically to $x^*(\mu)$. In general, the smaller μ, the smaller the attractive domain. The algorithm remains well defined as μ tends to zero. (The limiting case can be safely simulated in practice by using a very small value of μ.)

Note that feasible-direction methods can be made independent of the scaling of the search direction by appropriate re-scaling of the steplength α. We could therefore define the barrier search direction as

$$p = -Dr \tag{22}$$

for any $\mu \geq 0$. The "ideal" step would then be $\alpha = 1/\mu$.

The barrier search direction (22) with $\mu = 0$ in (20) is used in an algorithm proposed by Vanderbei, Meketon and Freedman [VMF8]. It has recently been discovered that this algorithm was first proposed by Dikin [Dik67]. From the above comments, we see that such an algorithm has no domain of quadratic convergence.

5.3 Upper bounds

The barrier transformation and the associated Newton search direction can be defined for linear programs with both upper and lower bounds on the variables, i.e., for problems of the form

$$\underset{x \in \Re^n}{\text{minimize}} \quad c^T x$$
$$\text{subject to} \quad Ax = b, \ \ l \leq x \leq u.$$

The subproblem analogous to (10) is

$$\underset{x \in \Re^n}{\text{minimize}} \quad c^T x - \mu \sum_{j=1}^n \ln(x_j - l_j) - \mu \sum_{j=1}^n \ln(u_j - x_j)$$
$$\text{subject to} \quad Ax = b.$$

The Hessian of the associated barrier function will be positive definite if at least one of l_j or u_j is finite for every j. In this case, the least-squares problem analogous to (20) is

$$\underset{\pi}{\text{minimize}} \ \|Dc - \mu \bar{D}e - DA^T\pi\|_2.$$

The matrices D and $\bar{D}$ are defined by $D = \text{diag}(\delta_j)$ and $\bar{D} = \text{diag}(\bar{\delta}_j)$, where

$$\delta_j = 1/(1/s_j^2 + 1/t_j^2)^{\frac{1}{2}} \quad \text{and} \quad \bar{\delta}_j = \delta_j(1/s_j - 1/t_j),$$

with $s_j = x_j - l_j$ and $t_j = u_j - x_j$.

For simplicity, the remainder of the discussion will assume that the bounds are of the simpler form $0 \le x_j \le \infty$.

6 Relationship with Karmarkar's Projective Method

In this section, we show the connection between the barrier and projective method.

In the projective method, the linear program is assumed to be of the special form

$$\begin{array}{ll} \underset{x \in \Re^n}{\text{minimize}} & c^T x \\ \text{subject to} & Cx = 0, \quad e^T x = 1, \quad x \ge 0. \end{array} \tag{23}$$

Let x_K^* denote a solution of (23). It is also assumed that

$$c^T x_K^* = 0 \tag{24}$$

and that $Ce = 0$.

Let π_C denote the solution of the least-squares problem

$$\underset{\pi}{\text{minimize}} \ \|Dc - DC^T\pi\|_2.$$

We also define

$$\begin{aligned} r_C &= Dc - DC^T\pi_C, \\ \mu_C &= x^T r_C. \end{aligned}$$

It was shown in Gill *et al.* [GMSTW86] that like the barrier function method, Karmarkar's projective method generates a sequence of iterates of the form

$$x_K \leftarrow x_K + \alpha_K p_K.$$

As in the barrier method, the choice of α_K is to some degree arbitrary, provided the new iterate is strictly feasible (i.e., interior). We have introduced a subscript to distinguish the step and search direction from that of the barrier algorithm. The following theorem states the relationship between the two methods. For a full discussion of this result, see [GMSTW86].

Theorem 6.1. *Suppose that the projective method and the barrier method are applied to problem (23), using the same initial point. If the barrier parameter is $\mu = \mu_C$, the search directions p and p_K are parallel. Further, if the steplengths satisfy $\alpha = \alpha_K\mu_C$, the iterates x and x_K are identical.*

Theorem 6.1 is an existence result, showing that a special case of the barrier method would follow the same path as the projective method. This does not mean that the barrier method *should* be specialized. For example, the value μ_C is an admissible barrier parameter only if it is positive. Note that μ_C *is* positive initially, if the starting point x_0 is a multiple of e. Furthermore, μ_C tends to zero as the iterates converge to x_K^*, and could therefore be a satisfactory choice for the barrier algorithm as the solution is approached.

Similarly, as the barrier method converges to a solution of the original problem for any suitable sequence of barrier parameters, θ will converge to λ_e, which is zero. This is consistent with the choice $\mu = \mu_C$, which gives $\theta = 0$ directly.

Recently, Monteiro and Adler [MA87] have shown that if the initial barrier parameter is chosen sufficiently large and then reduced at a sufficiently low rate, such an algorithm has a polynomial complexity that is better than that of the Karmarkar algorithm. It would appear, therefore, that there is nothing intrinsically preferable about the sequence of μ's implied by the Karmarkar algorithm.

7 Computational Aspects of the Barrier Method

The major computational effort of the barrier method is in solving the linear least-squares problem (20). Much research is currently under way on how best to solve sparse least-squares problems. We shall describe briefly some of the approaches being explored.

The solution of a linear least-squares problem (20) may be found from the so-called normal equations:

$$AD^2A^T\pi = AD^2c. \tag{25}$$

Since AD^2A^T is positive definite, one means of solving this system is by finding the Cholesky factorization R^TR, where R is upper triangular. Unfortunately, it is easy to construct examples where A is sparse but AD^2A^T (and hence R) is dense. For example, AD^2A^T is dense if A has just one dense column. One approach to this difficulty that has met with some success is to apply the preconditioned conjugate-gradient method, see [GMW81]. Suppose instead of the Cholesky factor R we find the factor $\bar{R}$, where $\bar{R}^T\bar{R} = \bar{A}D^2\bar{A}^T$ and $\bar{A}$ is close to A except that any fairly dense columns have been replaced by zeros. (Other approximations to A may also be desirable.) If $\bar{R}$ is used as the preconditioning matrix in a conjugate-gradient method, it can be shown that the number of conjugate-gradient iterations required is in theory no greater than the number of columns for which A and $\bar{A}$ differ.

Another approach that is suitable for dealing with a moderate number of dense columns is to partition A (and AD^2A^T) and use properties of the associated Schur complement.

7.1 Sparse Cholesky factors

The Cholesky factorization is usually found in two stages. It is well-known that the sparsity of $\bar{R}$ depends critically on the ordering of the rows and columns of $\bar{A}D^2\bar{A}^T$. Since we wish to retain symmetry only symmetric permutations are considered. A good ordering may be determined symbolically. This is referred to as the *analyze* phase. The sparsity structure of the factors may also be determined. A *numerical* phase is then performed to compute the actual factors. Since the numerical stability of the algorithm is independent of the permutation there is no need to alter the permutation in the numerical phase.

This feature of the Cholesky factorization adds greatly to its relative efficiency compared to the more general LU factorization required in the simplex algorithm. Not only is there no need to perform tests on the pivots in the numerical phase, but the knowledge that there can be no additional fill-in greatly simplifies the required data structures. The sparsity structure of $\bar{R}$ is independent of D and therefore remains constant for each iteration of the barrier algorithm. Consequently the analyze phase need be performed only once.

The usual procedure employed to perform the analyze phase is the minimum-degree algorithm. This is similar to the procedure used to form an LU factorization except it is now simplified, since we seek only diagonal pivots. Because the analyze phase need be performed only once, it may be worthwhile to invest more effort to find a better permutation than that resulting from the minimum-degree algorithm. One possibility is the minimum-fill algorithm. In this approach the pivot chosen at each stage of the factorization is the one that will result in the least increase of nonzero elements. It differs from the minimum-degree algorithm when a row that does not have the minimum number of elements has its nonzero elements in columns that match closely those in the other rows.

7.2 Storage

Having determined a good permutation for $\bar{A}D^2\bar{A}^T$, we still have a number of options on how to use the information. The most efficient approach in terms of storage is to store the nonzero elements in condensed format: the numerical values of nonzeros are stored in one array, and two integer arrays are used to hold pointers to the start of each column and the indices of the nonzeros in the columns. An alternative is to define an integer array containing the indices whose elements are used to construct a given element. This reduces the amount of indirect addressing required when performing the numerical phase of the factorization.

Much of the efficiency of the barrier approach stems from the efficiency of the latest techniques to factorize positive-definite matrices—in particular, the advantage that can be taken of the inherent stability of the factorization irrespective of the numerical values of the elements.

7.3 Ill-conditioning

Ill-conditioning in $\bar{A}D^2\bar{A}^T$ is inevitable if the number of elements of D that are small (compared to the remaining elements) is greater than $n - m$. This will be the case if an iterate is close to a large number of bounds (i.e., if the current x is close to a degenerate vertex). Consequently the choice of the initial point is quite crucial to the success of the method. If the solution is degenerate, $\bar{A}D^2\bar{A}^T$ will eventually become ill-conditioned. Unfortunately, most practical LP problems are degenerate. Because of the potential difficulties, attention has been given to preprocessing problems to remove redundant constraints and fixed variables.

It is sometimes the case that all feasible vertices are degenerate. This can cause difficulties in finding a feasible point (see next section). If $\bar{A}D^2\bar{A}^T$ is excessively ill-conditioned, the Cholesky factorization will break down. By modifying the diagonals of $\bar{A}D^2\bar{A}^T$ it is possible to avoid such a breakdown. However, such a modification needs to be done with care to avoid a loss of feasibility.

If an iterate is close to a small number of bounds (less than $n - m$), the Hessian is ill-conditioned and the search direction may be poor. Also, as a consequence of numerical error, the search direction may not be a descent direction for the barrier function. Under such circumstances it can be shown that the search direction is almost orthogonal to the constraint normals of those constraints in the vicinity of an iterate. In theory this is of no consequence if the constraints are indeed active at the solution (but it may still lead to embarrassing numerical errors). However, regardless of the accuracy of the computation, the search direction will be poor if even one of the constraints that is currently close to an iterate is not active at the solution.

Current research is directed at methods to regularize the definition and computation of p.

7.4 Finding an initial feasible point

In contrast to the simplex method, the barrier approach requires not just a feasible point but an *interior* feasible point. A similar Phase-1 procedure may be used, but instead of minimizing the sum of the infeasibilities relative to the bounds, it is necessary to minimize the infeasibilities in the general constraints. Provided an interior point exists, applying the barrier algorithm to the Phase-1 problem will yield a feasible point for a positive value of the barrier parameter. However, we cannot be assured that the feasible point is not close to the boundary of the feasible region.

An alternative approach is to use a composite objective function. For any given positive initial point x_0, we define $\xi_0 s = b - A x_0$ (with $\|s\| = 1$) and solve the modified linear program

$$\begin{aligned}
\underset{x,\xi}{\text{minimize}} \quad & \omega\, c^T x + \xi \\
\text{subject to} \quad & \begin{pmatrix} A & s \end{pmatrix} \begin{pmatrix} x \\ \xi \end{pmatrix} = b, \quad x \geq 0, \quad \xi \geq -1,
\end{aligned} \tag{26}$$

using the feasible starting point $x_0 > 0$, $\xi_0 = \|b - A x_0\|$. (Note that even if A is sparse, the additional column s in (26) will in general be *dense*.)

Assuming the original LP has a feasible solution and ω is sufficiently small, x^* will be a solution of the above LP. The lower bound on ξ is set to -1; otherwise a feasible point could be reached only as $\mu \to 0$. If the step α causes ξ to become negative, an appropriate shorter step is taken. When $\xi = 0$, a suitable feasible point has been found and the additional variable ξ may be dropped. The original linear program is presumed to be infeasible if the final ξ is positive for a sufficiently small value of μ.

In practice, a suitable value of ω is not necessarily known. This introduces a certain inefficiency if the initial value chosen is too large. If the initial ω does not produce a feasible point, its value can be decreased. It quickly becomes apparent whether decreasing ω is likely to yield a feasible point.

8 A Dual Algorithm

An alternative to solving LP is to apply the barrier algorithm to the dual problem,

$$\begin{aligned}
\underset{y \in \Re^m}{\text{minimize}} \quad & b^T y \\
\text{subject to} \quad & A^T y \geq -c.
\end{aligned}$$

The corresponding subproblem is

$$\underset{y \in \Re^m}{\text{minimize}} \quad b^T y - \mu \sum_{j=1}^{n} z_j \ln z_j,$$

where $z_j = c_j + a_j^T y$. Applying Newton's method leads to the following equations for determining the search direction p:

$$A D^{-2} A^T p = \mu(A x - b), \tag{27}$$

where $D = \operatorname{diag}(z_j)$ and $x_j = \mu / z_j$. The key point is that the dual approach leads to an *unconstrained* subproblem. It is therefore possible to consider search directions other than an accurate solution of (27).

Although we have transformed the problem to be unconstrained, we cannot start from an arbitrary point. The initial value of y must be dual feasible—i.e., the associated $z = c + A^T y$ must be (sufficiently) positive.

It is possible to rewrite (27) as a least-squares problem of the form

$$\min \| D(f - A^T p) \|_2.$$

It can be shown that $\lim_{\mu \to 0} D_D = \lim_{\mu \to 0} D_P$, where the subscript is introduced to distinguish between the diagonal matrices arising in the primal and dual methods. A consequence is that the condition of the least-squares problems to be solved in each approach is identical in the limit. To some extent this is regrettable, since it implies that the primal and dual algorithms have numerical difficulties on the same class of problems.

9 Alternative Barrier Functions

There are clearly choices of barrier function other than the logarithmic one used above. It may be that other choices lead to better conditioned subproblems in the case where LP has a degenerate solution. In place of (10), one can consider the subproblem

$$\begin{array}{ll} \underset{x \in \Re^n}{\text{minimize}} & c^T x + \mu \sum_{j=1}^{n} x_j \ln x_j \\ \text{subject to} & Ax = b, \end{array} \tag{28}$$

where the scalar μ ($\mu > 0$) is again specified for each subproblem. Erlander [Erl77] reviews problems of this kind and suggests Newton-type methods for their solution. Computational algorithms have been developed by Eriksson [Eri80,Eri81,Eri85] (see next section). This work preceded that of Karmarkar and should perhaps receive more recognition. Related work has been performed by Stewart [Ste80,Ste87], who considered the problem of finding a feasible point (which is essentially equivalent to an LP problem).

If a feasible-point descent method is applied as in Section 5, the Newton search direction and Lagrange-multiplier estimates satisfy the system

$$\begin{pmatrix} \mu D^{-1} & A^T \\ A & 0 \end{pmatrix} \begin{pmatrix} -p \\ \pi \end{pmatrix} = \begin{pmatrix} c + \mu v \\ 0 \end{pmatrix}$$

in place of (18), where $D = \mathrm{diag}(x_j)$ and v has components $v_j = 1 + \ln x_j$. A least-squares subproblem follows as before.

The entropy function is convex and (unlike the logarithmic barrier function) bounded below. Since its Hessian is μD^{-1} rather than μD^{-2}, the least-squares problems are likely to be better conditioned than those arising for the logarithmic barrier function. When LP is degenerate, this is certain to be the case in the neighborhood of the solution.

Again, the main computational part of the algorithm is solving a sparse least-squares problem. Research gains made on implementing the logarithmic barrier function are likely to be immediately applicable to the entropy approach.

9.1 Eriksson's algorithm

A significantly *different* algorithm was given in 1981 by Eriksson [Eri81] and further developed in [Eri85]. In place of (28), a primal subproblem is introduced that depends on the current estimate of x (x^k):

$$\text{Primal:} \quad \underset{x}{\text{minimize}} \quad c^T x - \mu \sum_{j=1}^{n} \left\{ x_j \ln(x_j/x_j^k) - (x_j - x_j^k) \right\} \tag{29}$$
$$\text{subject to} \quad Ax = b.$$

It is assumed that $x^k > 0$, but not that $Ax^k = b$. Instead, the *dual* of (29) is treated as an unconstrained problem in π:

$$\text{Dual:} \quad \underset{\pi}{\text{maximize}} \quad b^T \pi - \mu \sum_{j=1}^{n} x_j^k e^{-z_j/\mu}, \tag{30}$$

where $z = c - A^T \pi$. An inexact Newton method is applied to (30), with the central system of equations taking the form

$$ADA^T q = b - Ax^k. \tag{31}$$

This algorithm has many intriguing properties, and we believe it to be of great promise. For example, the matrix ADA^T will in general be better-conditioned than the usual AD^2A^T. Competitive computational results await implementation of a sparse preconditioner for (31), using techniques that have been applied to barrier algorithms elsewhere.

10 Summary

Since the renewed interest in nonlinear methods for linear programming occurred, there has been considerable computational testing of this class of algorithm. Although it is too early to draw complete conclusions, some properties of the approach are clear. If the Cholesky factors of AD^2A^T or $\bar{A}D^2\bar{A}^T$ are sparse (where $\bar{A}$ is some some sense similar to A), the methods, especially on very large problems, can outperform the simplex algorithm. However, we should point out that all published comparisons with the simplex method assume no prior knowledge of the problem is known. A difficulty with most of the new methods is that they are unable to make use of such information. In particular it is not known how to make use of a "good" starting point. (Eriksson's algorithm is an exception.) In contrast the simplex method is very flexible in this respect. For this reason we believe that the simplex method will remain the workhorse for the majority of existing applications, given practitioners' normal mode of operation: frequent restarts on slightly modified models. What the new methods do is to give practitioners a new tool that may be used to solve certain problems of a size that previously they may have considered intractable. Undoubtedly further research will extend the usefulness of the new methods.

References

[Ben62]   J. F. Benders. Partitioning procedures for solving mixed-variables programming problems, *Numerische Mathematik* 4, 238–252, 1962.

[Bla77]   R. G. Bland. New finite pivoting rules for the simplex method. *Mathematics of Operations Research*, 2, 103–107, 1977.

[Dan63]   G. B. Dantzig. *Linear Programming and Extensions*, Princeton University Press, Princeton, NJ, 1963.

[Dik67]   I. I. Dikin. Iterative solution of problems of linear and quadratic programming, *Soviet Math. Doklady* 8, 674–675, 1967.

[Eri80]   J. R. Eriksson. A note on solution of large sparse maximum entropy problems with linear equality constraints, *Mathematical Programming* 18, 146–154, 1980.

[Eri81]   J. R. Eriksson. Algorithms for entropy and mathematical programming, Ph.D. thesis, Department of Mathematics, Linköping University, Linköping, Sweden, 1981.

[Eri85]   J. R. Eriksson. An iterative primal-dual algorithm for linear programming, Report LiTH-MAT-R-1985-10, Department of Mathematics, Linköping University, Linköping, Sweden, 1985.

[Erl77]   S. Erlander. Entropy in linear programs—an approach to planning, Report LiTH-MAT-R-77-3, Department of Mathematics, Linköping University, Linköping, Sweden, 1977.

[Fia79]   A. V. Fiacco. Barrier methods for nonlinear programming, in A. Holzman, editor, *Operations Research Support Methodology*, pages 377–440, Marcel Dekker, New York, 1979.

[Fle81]   R. Fletcher. *Practical Methods of Optimization. Volume 2: Constrained Optimization*, John Wiley and Sons, Chichester and New York, 1981.

[FM68]   A. V. Fiacco and G. P. McCormick. *Nonlinear Programming: Sequential Unconstrained Minimization Techniques*, John Wiley and Sons, New York, 1968.

[FM84]   R. Fletcher and S. P. J. Matthews. Stable modification of explicit *LU* factors for simplex updates, *Mathematical Programming* 30, 267–284, 1984.

[Fri55]   K. R. Frisch. The logarithmic potential method of convex programming, University Institute of Economics, Oslo, Norway, 1955.

[FT88]   J. J. H. Forrest and J. A. Tomlin. Vector processing in simplex and interior methods for linear programming, Manuscript, Presented at the workshop on *Supercomputers and Large-Scale Optimization*, University of Minnesota, 1988.

[GMSTW86]   P. E. Gill, W. Murray, M. A. Saunders, J. A. Tomlin and M. H. Wright. On projected Newton barrier methods for linear programming and an equivalence to Karmarkar's projective method, *Mathematical Programming* 36, 183–209, 1986.

[GMSW84]   P. E. Gill, W. Murray, M. A. Saunders and M. H. Wright. Sparse matrix methods in optimization, *SIAM Journal on Scientific and Statistical Computing* 5, 562–589, 1984.

[GMSW88]   P. E. Gill, W. Murray, M. A. Saunders and M. H. Wright. A practical anti-cycling procedure for linear and nonlinear programming, Report SOL 88-4, Department of Operations Research, Stanford University, Stanford, CA, 1988.

[GMW81]   P. E. Gill, W. Murray, and M. H. Wright. *Practical Optimization.* Academic Press, London and New York, 1981.

[Har73]   P. M. J. Harris. Pivot selection methods of the Devex LP code. *Mathematical Programming,* 5, 1–28, 1973. Reprinted in *Mathematical Programming Study,* 4, 30–57, 1975.

[Jit78]   K. Jittorntrum. *Sequential Algorithms in Nonlinear Programming,* Ph.D. Thesis, Australian National University, Canberra, Australia, 1978.

[JO78]   K. Jittorntrum and M. R. Osborne. Trajectory analysis and extrapolation in barrier function methods, *Journal of Australian Mathematical Society Series B* 20, 352–369, 1978.

[JO80]   K. Jittorntrum and M. R. Osborne. A modified barrier function method with improved rate of convergence for degenerate problems, *Journal of Australian Mathematical Society Series B* 21, 305–329, 1980.

[Kar84]   N. Karmarkar. A new polynomial-time algorithm for linear programming, *Combinatorica* 4, 373–395, 1984.

[MA87]   R. D. C. Monteiro and I. Adler. An $O(n^3 L)$ primal-dual interior point algorithm for linear programming, Technical Report, Department of Industrial Engineering and Operations Research, University of California, Berkeley, CA, 1987.

[Mif72]   R. Mifflin. On the convergence of the logarithmic barrier function method, in F. Lootsma, editor, *Numerical Methods for Nonlinear Optimization,* pages 367–369, Academic Press, London, 1972.

[Mif75]   R. Mifflin. Convergence bounds for nonlinear programming algorithms, *Mathematical Programming* 8, 251–271, 1975.

[Mot51]   T. S. Motzkin. New techniques for linear inequalities and optimization, in proceedings of symposium on *Linear Inequalities and Programming,* Washington, DC, 1951.

[Mur69]   W. Murray. *Constrained Optimization,* Ph.D. Thesis, University of London, 1969.

[MW78]   W. Murray and M. H. Wright. Projected Lagrangian methods based on the trajectories of penalty and barrier functions, Report SOL 78-23, Department of Operations Research, Stanford University, Stanford, CA, 1978.

[Ste80]   G. W. Stewart. A new method for solving linear inequalities, Report TR-970, Department of Computer Science, University of Maryland, College Park, MD, 1980.

[Ste87] G. W. Stewart. An iterative method for solving linear inequalities, Report TR-1833, Department of Computer Science, University of Maryland, College Park, MD, 1987.

[VMF8] R. J. Vanderbei, M. S. Meketon and B. A. Freedman. A modification of Karmarkar's linear programming algorithm, *Algorithmica* 1, 395–407, 1986.

[Wri76] M. H. Wright. *Numerical Methods for Nonlinearly Constrained Optimization*, Ph.D. Thesis, Stanford University, Stanford, CA, 1976.

EXTENDED ABSTRACTS

RESOURCE CONSTRAINED ASSIGNMENT PROBLEMS

by

Ronny Aboudi & Kurt Jörnsten
Department of Science and Technology
Chr. Michelsen Institute, Norway

In many applications it is necessary to find a minimum weight assignment that satisfies one or several additional resource constraints. For example, consider the problem of assigning persons to jobs where each assignment utilizes at least two scarce resources and the resource utilization is dependent on the person and the type of task. A practical situation where the above might occur is a slaughter house where the "cutters" are assigned to different cut patterns. In this case the resources are the time, the cost and the productivity measured in terms of quality and amount of the end products.

In this paper we study the resource constrained assignment problem and derive several classes of valid inequaities based on the properties of the knapsack and assignment problems.

We also present an algorithm that uses both the linear programming and the Lagrangean relaxation of the original problem in order to solve the separation problem. Some computational experiments are also given.

THE RESOURCE CONSTRAINED SPANNING TREE PROBLEM:
ALTERNATIVE MODELLING AND ALGORITHMIC APPROACHES

by

Jaime Barceló
Dept. d´Investigacio Operativa i Estadistica
Facultat d´Informatica de Barcelona, Spain
&
Kurt Jörnsten
Department of Science and Technology
Chr. Michelsen Institute, Norway
&
Sakis Migdalas
Dept. of Mathematics
University of Linköping, Sweden

The problem of determining a minimal spanning tree subject to side
constraints, arises frequently as a subproblem of the general network
design problem, specially in the design of computer communication
networks and pipeline systems.

This problem can be modelled in different ways given that the inherent
structure of the side constraints may vary for different applications.
Our paper studies some of these modelling possibilities discussing its
advantages or disadvantages from an algorithmic point of view.

Two sets of modelling alternatives are included in our paper: (a)
formulating the problem as a particular case of discrete choice
network design problem, and (b) using an integer programming
formulation.

We focus our attention on these modelling alternatives because they
allow the use of a Lagrangean Relaxation approach, showing that some
of these formulations imbed combinatorial structures that can be
successfully exploited by Lagrangean Techniques.

This is particularly true when the side constraints are knapsack type
constraints, and then the problem includes two of such combinatorial
substructures: the three-like structure in the underlying network and
the knapsack or multiknapsack structure.

This is the case for the integer programming formulation where a first
algorithmic possibility is to try a direct polyhedral approach, adding
to the LP relaxation valid inequalities violated by the current LP
solution identified either from the knapsack type constraints (side
constraints) of from the subtour elimination constraints (tree
facets), also in some cases the LP solution gives some disconnected
components that lead to other valid inequalities.

After reviewing the previous work done by Jörnsten and Migdalas using a reformulation of the problem through the utilization of a "variable splitting technique", we study another way of exploiting the combinatorial substructure for algorithmic purposes in the framework of the Lagrangean Relaxation Techniques. We present a way of including the results from the polyhedral theory, already used with the direct polyhedral approach, in the lagrangian relaxation schemes.

Our paper describes a lagrangean relaxation algorithm that at each iteration expands the dual space by adding to the dual function a new valid inequality for the problem generated by the current partial solution. Examples are given to illustrate the procedure.

CONSTRUCTIVE DUAL METHODS FOR NON-LINEAR DISCRETE PROGRAMMING PROBLEMS

by

P. Bárcia & J.D. Coelho
Faculdade de Economia
Universidade Nova de Lisboa, Portugal

Recently a constructive duality theory for integer linear programming has been suggested, Bárcia (1985) and (1986). In this paper we generalize the previous theory for the case of non-linear discrete programming and present an algorithm for the case of quadratic 0-1 problems.

The algorithm is based on a modification of the ´variable splitting´ modelling technique, Jörnsten and Nasberg (1986), used as hinted in Bárcia and Jörnsten (1986), by relaxing a part of the integrality constraints in one of the ´variable splitting´ sub-problems.

The use of this modelling technique enables us to get an algorithm that is computationally tractable for some quadratic 0-1 problems.

In particular this approach allow us to solve an important class of quadratic assignment problems that arise when interaction generates some kind of benefit in a cost maximisation framework.

References

1) Bárcia, P. (1985) "The bound improving sequence algorithm", OR letters, Vol. 4, No. 1, pp. 27-30.

2) Bárcia, P. (1986) "Constructive dual methods for discrete programming", Discrete Applied Mathematics, forthcoming.

3) Bárcia, P.; Jörnsten, K. (1986) "Constructive duality and variable splitting: a method for structured integer programming", paper presented at the EURO VIII conference, Lisbon.

A DECOMPOSITION BASED PROCEDURE FOR PRODUCTION SCHEDULING IN JOB-SHOPS WITH NEGLIGIBLE SETUP TIMES

by

Ömer S. Benli

Bilkent University, Turkey

In short term planning of production in a discrete parts manufacturing system, a major problem is to decide on the maximum number of prespecified items that can be produced with limited shared resources, when all the items have a demand dependence in the form of bill-of-materials. The inventory carrying costs and the variation in the unit production costs from period to period become relatively unimportant in case of short planning horizons (such as one month, with periods taken to be one shift). When the total number of interacting items (such as raw materials, semifinished parts, subassemblies, finished goods, etc.) is in the order of thousands, with considerable number of shared resources (such as machines, work centres, etc.), the resulting model has a very large number of constraints and variables. Inclusion of the setup costs (or setup times), because of the introduced nonconvexity, makes the solution of the mathematical program infeasible from practical and computational viewpoint for short term planning purposes. Even the exclusion of the setup costs results in a large scale linear program.

In this paper, a decomposition procedure is presented for this large scale linear programming problem in which the master problem has only the resource constraints (thus reducing the number of constraints of the original problem by the number of items times the number of periods), and the subproblems by which the columns are generated, are one-pass procedures.

There are two types of constraints in the linear programming formulation of the problem: inventory balance equations and the resource constraints. Inventory balance equations are a set of linear homogeneous linear equations. Every feasible solution to this set can be expressed as a nonnegative linear combination of the extreme homogeneous solutions of this system of equations. Using this fact the original problem can be equivalently represented as a linear program. The master problem thus obtained has a very large number of variables but the number of constraints is equal to the number of resource constraints of the original problem. Applying restriction solution strategy to this problem results in subproblems. The subproblems, after proper manipulation, can be solved by a one-pass procedure.

The objective function used in this presentation is basically to maximize the throughput for a given product mix. In a follow-up study a different objective function, namely minimizing the makespan for a given demand vector, was used. Furthermore in that study experimental runs were made with various "surrogate" objective functions. Interesting results were obtained with the objective function which forces items to be processed in one setup. Analyses was also carried out with different priorities assigned on items and different product tree structures.

The algorithm was coded in PL/1 and implemented on Burroughs System Serial 1056, A9/F. Currently the research is under way to implement the algorithm in a real production environment with an improved computer code to be executed on microcomputers.

AN EXAMPLE OF A DECLARATIVE APPROACH TO MODEL CREATION

by

Iain Buchanan
Computer Science Department
Strathclyde University, Scotland
&
K.I.M. McKinnon
Department of Mathematics
Edinburgh University, Scotland

Model creation tools are often closely coupled to the target
mathematical structures. In the case of constrained optimisation,
model creation is normally via the concepts of matrix or linear
algebra, and thus perhaps only indirectly related to the objects and
concepts of the problem domain. A translation from problem structure
to mathematical structure is necessary.

The creation of a mathematical model of complex domain is a highly
skilled task, involving an understanding of the domain, the design
choices in the formulation process and the implications of these for
run-time performance and solution quality. If a large number of
problems are to be solved within the same domain by a person who is
not an expert in mathematical modelling, it makes sense to consolidate
these modelling skills into a program. The user should be able to
define his model in a language which refers to objects in his problem
domain, and the program should translate this to a form compatible
with the internal structure of the solver. This paper describes the
approach for a power system scheduling domain.

The user specifies the system to be solved by defining sets of facts
about objects in the system, and relations between these objects,
using a declarative language (Prolog). Objects in the domain are
typed, e.g. as thermal sets, catchments. Objects can have associated
properties, functions and initial/final conditions. Properties
describe (potentially) time-variant data, e.g. the operating
characteristics of a set. Functions are used to describe or constrain
the behaviour of objects over time, e.g. a user lower bound on
generator output or the level of demand. Properties/functions can be
defined at a 'raw' level or may be composed by taking portions out of
other properties/functions. Some association of objects is necessary
(e.g. to produce a network topology). The model can be built from
components in an arbitrary manner and the system incorporates
substantial model management and memory management facilities.

The declaration is compiled to a temporary internal form which is
interrogated by the program which generates the internal form for the

solver.   This program is aware of the domain, the capabilities of the solver, and good modelling practice.   In the power scheduling study, the model contains non-convexities and nonlinearities.   Non-convexities in the model are removed by adopting a bang-bang policy and a Lagrangian technique used to transfer nonlinearities from constraints to the objective function, which is then approximated by a piecewise linear (PWL) function.   A sequence of these approximations is solved using LPSolve, a fast revised simplex implementation for problems with linear constraints and PWL separable objective functions.

OPTIMAL SOLUTION OF THE LOCAL DELIVERY PROBLEM
THROUGH MINIMUM K-TREES

by

Marshall L. Fisher
Department of Decision Sciences, The Wharton School
University of Philadelphia, USA

The local delivery problem has been extensively studied and is a
classic in the field of vehicle routing and scheduling. This talk
will describe a recently developed optimization algorithm for this
problem which exploits a connection between the local delivery problem
and minimum K-trees. I will first set the stage for this discussion
by reviewing the enormous activity in the vehicle routing area that
has occured during the last decade in both academia and industry. I
will focus in particular on projects in which I have been involved at
Air Products and Chemicals, DuPont and Exxon that produced implemented
routing models and algorithms with significant economic benefits.
Altough all of these algorithms are heuristics (as has been the focus
of almost all past research in this area), I will indicate why the
time is now ripe to consider optimization approaches.

Attention will then be turned to optimal solution of the local
delivery problem through K-trees. Given a graph with n+1 nodes, a K-
tree is a set of n+K arcs that span the graph. In the local delivery
problem we must accomplish n customer deliveries using a fleet of K
vehicles stationed at a central depot. Each vehicle has a fixed
capacity, customer orders have a specified size, and the cost of
travel between any two points in the problem is given. We are
required to assign each customer to one of the vehicles and sequence
the order in which the customers are delivered so as to minimize total
travel cost without exceeding vehicle capacity constraints.

We can define a complete graph with node set equal to the customers
plus the depot and edge weights given by the direct travel costs. The
local delivery problem is then equivalent to the problem of finding a
minimum K-tree with degree 2K at the depot node, degree 2 at each
customer node and satisfying a set of side constraints that impose the
vehicle capacity constraints.

A Lagrangian problem is obtained by dualizing the customer degree and
vehicle capacity constraints. The Lagrangian problem is a degree
constrained minimum K-tree problem for which I will exhibit an
efficient polynomial algorithm. The lower bound provided by the
Lagrangian problem is used with a branching rule to obtain an
optimization algorithm for which I´ll report computational experience
on problems with up to 200 customers.

AMPL: A MATHEMATICAL PROGRAMMING LANGUAGE

Robert Fourer
Department of Industrial Engineering
Northwestern University, USA
and
AT&T Bell Laboratories, USA
&
David M. Gay
AT&T Bell Laboratories, USA
&
Brian W. Kernighan
AT&T Bell Laboratories, USA

Practical large-scale mathematical programming involves more than just the minimization or maximization of an objective function subject to constraint equations and inequalities. Considerable effort must be expended to correctly formulate the underlying model, and to generate the data structures required by an optimizing algorithm.

These pre-optimization steps are problematical because people and algorithms deal with mathematical programs in very different ways. Traditionally, the difficult work of translation from "modeler's form" to "algorithm's form" is divided between human and computer: someone studies the modeler's form and writes a computer program, a computer compiles and executes the program, and the program writes out the algorithm's form. Such an arrangement is often costly and error-prone, particularly because it involves a program whose output is ill-suited to debugging.

Many of the difficulties of translation can be circumvented by use of a computer modeling language for mathematical programming. A modeling language expresses the modeler's form in a way that permits it to be interpreted directly by a computer system. Thus the translation to the algorithm's form is performed entirely by computer, and the intermediate computer-programming stage is avoided.

We describe in this paper the design and implementation of AMPL, a new modeling language for mathematical programming. AMPL is notable for the generality of its syntax, and for the similarity of its expressions to the algebraic notation customarily used by the

modeler's form. It offers a broad variety of sets and set operations,
as well as quite general logical expressions.

We intend AMPL to be able to express arbitrary mathematical
programming problems, including ones that incorporate nonlinear
expressions or discrete variables. However, our initial
implementation is restricted to linear expressions in continuous
variables. Thus AMPL is introduced by means of a simple linear
programming example; subsequent sections examine major aspects of the
language's design in more detail, with reference to three much more
complex linear programs. We also describe the implementation of the
language translator, and present a standard data format. Finally, we
compare AMPL to other modeling languages for linear programming.

MULTIPERIOD LINEAR STOCHASTIC PROGRAMMING AND A FORESTRY APPLICATION

by

Gus Gassmann

School of Business Administration

Dalhousie University, Canada

A new algorithm is given to solve the multiperiod stochastic linear programming problem

$$\min_{x_1} \{c_1 x_1 + E_{\xi_2} [\min_{x_2} (c_2 x_2 + E_{\xi_3} (\min_{x_3} c_3 x_3 + \ldots + E_{\xi_T} \min_{x_T} c_T x_T))]\}$$

$$\begin{aligned}
\text{s.t. } A_1 x_1 &= b_1 \\
B_2 x_1 + A_2 x_2 &= \xi_2 \\
B_3 x_2 + A_3 x_3 &= \xi_3 \qquad\qquad (1) \\
&\;\;\vdots \\
B_T x_{T-1} + A_T x_T &= \xi_T
\end{aligned}$$

$$x_t \geq 0, \quad t=1,\ldots,T.$$

All random elements are assumed discretely and finitely distributed. Like an earlier algorithm by Birge [1], the present method is based on the principle of nested decomposition, working directly on the primal problem. The new procedure is more flexible and accomodates stochastic cost vectors as well as stochastic constraint matrices.

Problem (1) can be decomposed into a collection of subproblems of the form

min

$$c_t x_t^{h_t} + v_t^{h_t}$$

s.t.

$$A_t x_t^{h_t} = \xi_t - B_{t+1} x_{t-1}^{h_t} \qquad (2.1)$$

$$\sum_{k=1}^{K_{t+1}} p_{t+1}^k \pi_k^i B_{t+1} x_t^{h_t} + v_t^{h_t} \geq \sum_{k=1}^{K_{t+1}} p_{t+1}^k [\pi_k^i \xi_{t+1}^k + \lambda_k^i l_{t+1} - \mu_k^i u_{t+1}], \quad i=1,\ldots,I(h_t) \quad (2.2)$$

$$\sigma_{k(j)}^j B_t x_t^{h_t} \geq \sigma_{k(j)}^j [\xi_{t+1}^{k(j)} + \lambda_{k(j)}^j l_{t+1} - \mu_{k(j)}^j u_{t+1}], \quad j=1,\ldots,J(h_t) \qquad (2.3)$$

$$l_t \leq x_t^{h_t} \leq u_t$$

Here k labels the nodes in the decision tree, $a(k)$ denotes the immediate predecessor of k, $x_{t-1}^{a(k)}$ is the current solution of the ancestor problem, and (2.2) and (2.3) are, respectively, optimality and feasibility cuts which give a partial description (lower bound) of the value function at node k.

If the constraints matrices A_t do not depend on the realizations ξ_t, then many of the subproblems (2) will share the same constraint coefficients and hence the same bases. This can be exploited in a 'trickling-down' scheme first suggested by Wets [3], which solves many subproblems simultaneously.

Various sequencing protocols can be given which prescribe the order in which the subproblems are to be solved. This is so because at each step it is possible to generate primal information, which is passed to the descendant problems as a new right hand side, or to pass dual information back to the ancestor in the form of a cut. Numerical results on some test problems from the Ho and Loute [2] set show that the algorithm may perform significantly better than MINOS.

The algorithm is then applied to a problem from forest management, where an optimal harvest policy has to be determined in the face of random destruction rates due to forest fires and other environmental hazards. The standard Type I forestry model, which is ordinarily used in conjunction with a scenario analysis of the problem, is shown to be deficient in some key aspects. Reformulating the problem as a stochastic program with recourse gives new insights and leads to an implementable harvest policy.

References

1) J.R. Birge, "Decomposition and partitioning methods for multistage stochastic linear programs", <u>Operations Research 33 (1985)</u> 989-1007.

2) J.K. Ho and E. Loute, "A set of staircase linear programming test problems", <u>Mathematical Programming 20 (1981)</u> 245-250.

3) R.J-B. Wets, "Large scale linear programming techniques in stochastic programming", in Yu. Ermoliev and R.J-B. Wets, eds., <u>Numerical Methods in Stochastic Optimization</u> (Springer Lecture Notes (to appear)).

A MULTI-PERIOD NETWORK DESIGN PROBLEM: MODEL AND SOLUTION TECHNIQUES

by

Åsa Hallefjord
Department of Science and Technology
Chr. Michelsen Institute, Norway

We give a brief description of a part of a project initiated in the spring of 1985 at the Chr. Michelsen Institute. The aim was to develop a long term planning model for the sequencing of petroleum production activities on the Norwegian shelf. The emphasis of the model is on the development of transport systems for petroleum products from the producing fields to gas customers and oil terminals onshore.

The following questions should be answered by the model:

- Which new fields should be put into production?
- When should the selected fields start to produce?
- Which means of transport should be chosen for the new fields?
- Should the means of transport for producing fields be changed?
- In what order should the selected transport modules be developed?
- How should the constructed transport system be utilized?

The purpose is to analyze different scenarios given the production possibilities and the demand for oil and gas.

The input to the model should consist of the same data that are otherwise used for decision making on these matters. Essentially, the data are the production and the investment profiles for fields, the investment profiles for pipelines and loading buoys (some of which may have limited lifetime), the variable transport costs, the price profiles for oil and gas, the upper and lower limits for the demand of each customer (terminal or gas consumer), a budget for each time period, the capacities in existing transport modules and a suggested topology for a maximal transport network.

A mathematical programming model was formulated to capture the most essential parts of the problem setting. The model has as the major component a directed network. This network actually consists of two subnetworks, one for oil and one for gas. These subnetworks are connected to each other at the petroleum fields only. The basic idea of the model is that the network can be extended over time (within a given planning horizon), as new fields are brought into production. Therefore, the model can be seen as a multiperiod network design problem with some nontrivial extensions.

The objective is to maximize the net present value of the scenario, expressed as the difference between the revenue from sale of oil and gas, and the costs connected with starting and operating the fields and the links.

The model is a large mixed-integer programming model. The solution strategy chosen was to combine a number of heuristics. Some are general heuristics for 0-1 programming problems, others are specially designed to exploit the network design structure.

The following techniques are used in combination:

Rounding of LP solution - Problem specific rounding procedure. Preprocessing - Valid inequalities are identified and added to strengthen the LP relaxation. Exchange heuristic - A problem specific procedure with 2-changes. The pivot and complement heuristic - The heuristic suggested by Balas and Martin, with some modifications. the ideal column heuristic - As developed by Martin and Sweeney, to reduce the size of the branch-and-bound tree.

The heuristics were incorporated into the ZOOM/XMP system, developed by Roy Marsten.

Preliminary results are very promising - a solution within a few percent from optimality is usually obtained within reasonable time. It is especially the preprocessing and the ideal column heuristic that have performed well.

FINITE-DIMENSIONAL VARIATIONAL AND QUASIVARIATIONAL INEQUALITIES: ALGORITHMIC DEVELOPMENTS AND APPLICATIONS IN SOCIO-ECONOMIC PLANNING

by

Patrick T. Harker
Department of Decision Sciences, The Wharton School
University of Pennsylvania, USA

Over the past several years the finite-dimensional variational inequality problem has been well studied from the perspectives of computation, sensitivity analysis and application. This paper will briefly review the variational inequality literature and will also discuss the finite dimensional quasivariational inequality problem from the above mentioned perspectives. In particular, three recent results will be discussed in detail. First, an acceleration step for the nonlinear Jacobi and projection algorithms will be presented along with empirical results. Second, a special case of the quasivariational inequality problem which arises in pseudo-Nash or social equilibria games is studied in detail. Characterizations of the solutions of such problems and their relationships with variational inequality solution are presented along with some results on the stability and sensitivity of these solutions. Finally, the theory and empirical validation of a restricted simplicial decomposition algorithm for large-scale, linearly-constrained variational inequalities will be described.

The talk concludes with a discussion of the ongoing applications of these results to the modelling of urban mass transit privatization in the United States, capacity planning in telecommunications networks, new product location, marketing distribution channels, and freight transport planning, along with thoughts on future extensions and applications of these methods.

STOCHASTIC EQUILIBRIUM PROGRAMMING FOR DYNAMIC OLIGOPOLISTIC MARKETS

by

A. Haurie
Marketing
Ecole des Hautes Etudes Commerciales, Canada
&
Y. Smeers
CORE
Université Catholique de Louvain, Belgium
&
G. Zaccour
Marketing
Ecole des Hautes
Etudes Commerciales, Canada

The aim of this paper is to clarify the relationship between the stochastic programming and dynamic programming approaches for the modelling of dynamic equilibria in a class of uncertain systems.

Often the modelling of economic systems leads to the consideration of a multiplicity of agents or players who are competing on an imperfect market. This is particularly the case when one tries to model the energy markets (e.g. the electricity market in the Northeast American States and Canadian Provinces, the gas market in Europe, the world oil Market etc.). The modelers involved in the development of such models have the possibility to extend the stochastic programming approach by replacing the single optimization criterion with an equilibrium computation which is performed by using recent advances in the numerical treatment of variational inequalities. We call this approach the stochastic equilibrium-programming approach. Since this approach deals with a game-theoretic concept, viz. the Nash equilibrium solution, in a dynamic setting, there should be a relationship with the dynamic theory of games initiated by R. Isaacs, and more recently surveyed by Basar and Olsder. For deterministic dynamic games one distinguishes between the Open-Loop and Closed-Loop equilibrium concepts while the theory of stochastic dynamic games is essentially based on the dynamic programming approach yielding a Closed-Loop equilibrium. Usually the Closed-Loop equilibria are very hard to compute.

In this paper we show that the stochastic equilibrium-programming approach deals with a particular class of strategies which we call S-adapted Open-Loop strategies. In the case of a single player system it can be easily shown that, for the class of systems considered, any closed-loop strategy has a "representation" through an S-adapted Open-Loop strategy. This means that the stochastic programming approach gives the same solution (in terms of the optimal value for the performance criterion) as the dynamic programming approach. In the case of a m-Player system, the stochastic equilibrium-programming approach leads to an interesting solution concept which is halfway between the completely adaptive feedback or Closed-Loop equilibrium solution and the completely nonadaptive Open-Loop. Although this equilibrium is not subgame perfect à la Selten, it can be a useful representation of the outcome of supply and exchange contracts between energy producing and energy consuming countries over a long time horizon and under uncertainty. These contracts often reflect the competition between the possible suppliers and also include provisions under which the contracts will adapt to random modifications of the economic environment.

The paper is organized as follows; In section 2 we present the class of systems considered, in the simplified framework of a two-stage dynamical structure which nevertheless retains the essential ingredients of a dynamic game. In section 3 we first deal with the single-player case and we show that the stochastic programming approach is "locally" equivalent to the dynamic programming approach. In section 4 we define the equilibrium concept and we show that, in the class of S-adapted Open-Loop strategies the characterization of an equilibrium is obtained through the stochastic equilibrium-programming approach. In section 5 we show how this approach can be used for the modelling of a dynamic oligopoly model à la Cournot, with investment activities and random perturbations on the demand laws. A multi-stage model representing the European gas market is considered. The players are the producers (Algeria, Holland, Norway and USSR). A numerical illustration is given. In conclusion we discuss the appropriateness of this solution concept in the energy modelling area and we briefly discuss the possibility to extend the approach to an asymmetrical information structure (e.g. Stackelberg equilibrium).

A DYNAMIC APPROACH TO OLIGOPOLISTIC MARKET EQUILIBRIUM

by

Adi Ben-Israel
Deptartment of Mathematical Sciences
University of Delaware, USA
&
Sjur D. Flåm
Institute of Economics
University of Bergen, Norway

We provide an algorithm for computing Cournot-Nash equilibria in
multi-commodity markets involving finitely many producers. The
algorithm amounts to follow a certain dynamical system all the way to
its steady state which happens to be a non-cooperative equilibrium.
The dynamics arise quite naturally as follows: Let each producer
continuously adjust his planned production, if desired, as a response
to the current aggregate supply. In doing so he is completely guided
by myopic profit considerations. We show, under broad hypothesis on
the market structure, that this adjustment process is globally,
asymptotically convergent to a Nash equilibrium.

Key words Non-cooperative Games, Cournot-Nash Equilibria, Oligopoly,
Variational Inequalities, Lyapunov Functions, Differential Indusions.

ESTIMATED PARAMETERS IN MATHEMATICAL PROGRAMMING:
MODELLING AND STATISTICAL ISSUES

by

Alan J. King
Department of Mathematics
University of British Columbia, Canada

Frequently it happens that the parameters needed to define a
mathematical program are not precisely known, but can be estimated by
sampling or derived from historical data. To be honest, the modeller
should treat such parameters as uncertain. This is simple to state
but raises profound modelling, algorithmic and analytical issues.

Each piece of data represents one possible realization of the unknown
parameter and thus yields a unique evaluation of the performance of a
given decision as measured by the objective. We assert that in order
to arrive at a decision that performs well for the full range of
samples present in the data, one should seek the decision that
optimizes the average performance. This is a stochastic optimization
problem. The Lake Balaton Eutrophication model [1] is a significant
example of this construction.

New techniques are available to estimate the uncertainty residing in
the solutions to such problems [2]. The crucial result is that the
asymptotic distribution of the solutions (which describes the residual
uncertainty in the asymptotic sense as the number of data points grows
large) may be obtained numerically by solving a certain random
quadratic program. Thus useful estimates of critical quantities such
as asymptotic bias and variance are within our capabilities.

[1] A.J.King, R.T.Rockafellar, L.Somlyody and R.J-B.Wets "Lake
eutrophication management: the Lake Balaton study", in: Numerical
Techniques for Stochastic Optimization, Yu.Ermoliev and R.J-B.Wets,
eds., Springer, 1987.

[2] A.J.King, Asymptotic Behaviour of Solutions in Stochastic
Optimization: Nonsmooth Analysis and the Derivation of Non-normal
Limit Distributions. Dissertation, University of Washington, 1986.

MODELLING FOR PARALLEL OPTIMIZATION

by

Robert R. Meyer
Computer Sciences Department and Mathematics Research Center
University of Wisconsin-Madison, USA

As a result of the development of both research and commercial
multiprocessors and multi-computers capable of implementing parallel
algorithms for optimization, there has been a renewed interest in both
theoretical and computational aspects of such techniques. From a
modelling viewpoint, it becomes important to formulate the problem in
such a manner that it can be decomposed into quasi-independent
subproblems that are suited to the computer architecture on which it
will be solved. One key issue in this regard is the granularity of
the computation, i.e. the sizes of the ´pieces´ that will be dealt
with in parallel. For example, in a traffic assignment problem, one
has the option of defining commodities (and hence subproblems) by
origin-destination pair og by origin. The latter approach is less
traditional, but gives rise to larger sub-problems that are better
suited to a computing environment in which communication between
processors is relatively expensive. Thus, one method of dealing with
granularity is to seek to modify an existing model in such a way as to
facilitate the decomposition of the problem into subproblems of
appropriate size. A related issue is whether one will initialize the
decomposition process by using intrinsic structures available a priori
from the model (such as commodities or time periods) or exploit the
previous solution of a related problem (which may, for example, allow
a tentative geographic decomposition of the area covered by the
model), or employ a heuristic splitting procedure based on the current
problem data and then dynamically modify the structure of the
decomposition as needed during the course of the solution process. It
is clear that in order to reduce the amount of information exchange
and to accelerate the solution process, it is desirable to partition
the model in a manner that is suited to the machine architecture and
reflects as much as possible a partition related to an optimal
solution. These topics will be addressed in the context of research
on large-scale nonlinear and generalized networks. We will describe
computational experience with two parallel computing systems at the
Computer Sciences Department of the University of Wisconsin: the
CRYSTAL multicomputer, a loosely-coupled token-ring network of 20 VAX
11/750´s, and the Sequent Balance 21000, a commercial shared-memory
multiprocessor with 8 processors.

LONG-TERM HYDRO-THERMAL COORDINATION OF ELECTRICITY GENERATION
THROUGH MULTICOMMODITY NETWORK FLOWS

by

Narcis Nabona

Department of Operations Research

Universitat Politecnica de Catalonia, Spain

Besides classical methods, such as dynamic programming, network flow
techniques have been extensively used to solve the problem of short
term hydro-thermal coordination of electricity generation. Nonlinear
flows in a replicated network, can adequately model the hydraulic
generation of a system of reservoirs, which complements thermal
generations in satisfying known forecasts of each hourly demand. Side
constraints or penalty functions must be used, at each time interval,
to prevent excessive or insufficient hydro-power generation.

Long-term hydro-thermal coordination differs from short-term in that
the demand for electricity, the availability of thermal power plant
and water inflows in head reservoirs are not deterministic, but only
known as probability distributions. Probabilistic demand and
probabilistic availability of thermal plant can be adequately modeled
through functions of probabilistic cost with respect to hydro
generation, found for each time interval, by a special program using
well known techniques. There remains the problem of finding the set
of hydro generations at each long term interval that, while satisfying
network constraints, minimizes the sum over the time intervals, of the
probabilistic cost functions of hydro generations. However, the
difficulty still is dealing with probabilistic water inflows.

Special codes of dynamic programming taking into account
simultaneously several possibilities of probable flow at each time
interval, can be used to solve the long term hydro-thermal
coordination. These codes find a policy of variation of water level
at each reservoir that satisfies network constraints simultaneously
for any set of probable inflows considered, and minimizes a weighted
sum, for all considered probable inflows, of the summations of

probabilistic cost functions at each interval. Probabilities of occurence of considered water inflows can be used as weights. Although this solution does not take into account all possible circumstances of inflow in a wide geographical area and throughout a long period of time, it is a quite sensible rule to go by.

In this work the long term problem is solved with multicommodity network flows. Considering the cummulative property distributions of inflows at each interval, the minimum inflows at each time interval (that with 1.0 probability) are taken to be commodity #1. Commodity #2 are the inflows in excess of the minimum that have, for instance 0.8 probability. Commodity #3 could be the inflows in excess of those with 0.8 probability, with 0.6 probability, and so on until the whole spectrum of possible inflows to be considered is adequately covered. A linearized objective function and a nonlinear version have been tried with real examples, and the results are reported.

The multicommodity algorithm coded is the primal partitioning one and a non standard initial set of basic feasible spanning trees has been also tried and compared to the standard initial solution with artificial arcs. An analysis is made of the computational requirements of that type of solution.

EQUILIBRATION OPERATORS FOR THE
SOLUTION OF CONSTRAINED MATRIX PROBLEM

by

A. Nagurney and A.G. Robinson
School of Management
University of Massachusetts, USA

The problem of determining a set of coefficients of a matrix which
collectively satisfy certain constraints has become known as the
Constrained Matrix Problem. It has been widely studied because of its
frequent appearance as a "core" problem in diverse applications. These
include estimation of input-output tables in the regional sciences, of
contingency tables in statistics, origin-destination flows in traffic
analysis, and social-national accounts in economics.

In its most general form, the problem is to compute the best estimate
of an unknown matrix, given some information to constrain the solution
set. This might include row and column sums, totals of subsets of
matrix elements, and bounds on the individual entries. In the
Euclidean norm the problem becomes that of minimizing a strictly
convex quadratic function subject to (amongst others) linear
constraints of the transportation type.

Recent algorithmic advances for the constrained matrix problem (cf.
Cottle et.al. NRLQ 33, 55-76 (1986) and references therein) have
renewed interest in quadratic programming over the transportation
polytope with bounded variables.

In this paper we consider the general constrained matrix problem, with
a general positive definite quadratic weighting matrix (rather than a
diagonal one), with upper bounds on the individual matrix entries, and
additional linear constraints to enhance the modelling aspects.

The computational procedure we porpose is a decomposition scheme which
resolves the main problem into a series of equilibrium subproblems of
three types, which we call the row, column, and cut-set forms. We

introduce equilibration operators for each of the three subproblem types, and embed the iterative solution of these in the price decentralization scheme of Pang, JOTA <u>49</u>, 107-134 (1986).

Equilibration operators were first introduced by Dafermos and Sparrow, JNBS <u>73</u>b, 91-118 (1969) for the traffic network assignment problem and extended to the framework of the spatial price equilibrium problem (where demands and supplies are <u>elastic</u>) by Nagurney, JRS <u>27</u>, 55-76, (1987) ORLetters (1987).

We provide recent computational experiences and discuss the relative strength of the operators. The computational results strongly suggest that the solution of the subproblems using a diagonal quadratic matrix and an exact algorithm, rather than an iterative one, is more efficient.

A GENERAL DYNAMIC NETWORK SPATIAL PRICE EQUILIBRIUM MODEL WITH GAINS AND LOSSES

by

Anna Nagurney
School of Management
University of Massachusetts, USA

The spatial price equilibrium models of Samuelson, and Takayama and Judge have provided the basic framework for the study of a variety of applications in the fields of agriculture, regional science, and energy markets. The central issue in such studies is the computation of the equilibrium regional production, consumption, and interregional commodity flow patterns.

Although Takayama and Judge formulated spatial price equilibrium models which were temporal, most of the recent advances in model formulation and algorithm development in a general setting (utilizing either a complementarity or variational inequality approach) have considered exclusively static problems.

Granted, as early as 1957, Samuelson noted that temporal models can be viewed as static models if the associations of carry-over costs between time periods with transport costs, and time periods with regions are made. However, direct replication of existing static models over time may not adequately address such important issues as inventorying at supply and demand markets and backordering. Furthermore, the situations of perishability, thefts, and losses of commodities, as well as, accretion over time, is not handled in this framework. Finally, it is not clear, in the absence of rigorous testing, whether or not the solution of such large-scale temporal problems is computationally feasible.

In this paper we consider the general (dynamic) finite horizon spatial price equilibrium problem with gains and losses over discrete time periods. The supply price of the commodity at any supply market in any given time period may depend upon, in general, the supply of the commodity at every supply market in every time period. Similarly, the demand price of the commodity in any given time period may depend upon the demand of the commodity at every demand market in every time period. The inventorying cost of the commodity at any supply market, at any demand market, as well as, the backordering cost at any demand market between two time periods, may, in general, depend upon the quantities inventoried at every supply and every demand market between every pair of successive time periods, and the quantities of the commodity shipped between every pair of supply and demand markets within every time period. The transportation cost of shipping the

commodity between any pair of supply and demand markets within any
time period, in turn, may, in general, depend upon the quantities of
the commodity shipped between every pair of supply and demand markets
within every time period, the quantities inventoried at every supply
and every demand market between every pair of successive time periods,
and the quantities backordered at every demand market between two
periods.

For this problem, we introduce a general dynamic spatial price network
equilibrium model to handle gains and losses through the use of arc
multipliers. This framework extends the dynamic model of Nagurney and
Aronson (1986) and allows for a more realistic representation of for
example, agricultural markets for perishable commodities and financial
markets with the characteristic credit multipliers. We then define
the governing equilibrium conditions with the incorporation of
multipliers and give alternative variational inequality formulations
of the problem over Cartesian products of sets. This model contains,
as special cases, many of the static and temporal models treated
earlier and is based crucially on the visualization of the problem as
a multiperiod network.

We then propose a Gauss-Seidel serial linearization decomposition
scheme by demand markets in time for the computation of the
equilibrium. For the embedded mathematical programming problem we
introduce a new equilibration operator to handle gains and losses. We
first provide computational experience for the equilibration operator
for dynamic problems which can be formulated as equivalent
optimization problems. We then embed the equilibration operator in
the Gauss-Seidel scheme and provide computational experience on a
variety of general examples to demonstrate the efficiency of the
method. The computers utilized are the CDC CYBER 830 at the
University of Massachusetts, Amherst, and the IBM 3090 at the Cornell
National Supercomputer Facility.

INCORPORATING THE CONCEPT OF INTERNAL RATE OF RETURN IN LINEAR AND INTEGER PROGRAMMING MODELS

by

Robert M. Nauss
School of Business Administration
University of Missouri-St.Louis, USA

The use of an internal rate of return (IRR) measure is common in
financial problems. Over the past decade the measure has been
enjoying an increased usage in the new municipal debt issue market.
Nauss (Management Science, July 1986) developed a procedure to
minimize the IRR for competitive bids for new issues of municipal
debt. While this problem was generally viewed as being a nonlinear
integer program, it was shown that the problem could be linearized so
that an integer linear program resulted. The linearization is
possible because there is only one change in the sign of the cash
flows over time. This assures that only one real root for the IRR
exists. A special purpose branch and bound algorithm was developed to
solve the problem in a matter of seconds so that it could be used in
actual applications.

In this paper we extend the use of IRR to other particular classes of
problems which have only one real root for IRR. Both linear and
integer programming examples will be presented. IRR may be used as a
criterion to be optimized. It may also be used in a constraint which
speficies that any solution generated must have an IRR less than or
greater than some specified value. Or it may be used in complex
multiple-phase problems where no particular value of IRR is specified,
but where a solution in one phase must have an IRR greater than or
less than an IRR for a solution in another phase. Applications of
each of these cases are described. General modelling approaches are
developed for each case, and where necessary, algorithmic
modifications to existing solution approaches are described. In the
final section we explore relaxing the condition which requires that
only one real root exist.

PROCEDURES FOR SOLVING BOTTLENECK GENERALIZED ASSIGNMENT PROBLEMS

by

Alan W. Neebe
School of Business Administration
University of North Carolina, USA
&
Joseph B. Mazzola
Fuqua School of Business
Duke University, USA

We discuss bottleneck (or minimax) versions of the generalized assignment problem. The basic problem involves the assignment of a number of jobs to a number of agents such that each job is performed by a unique agent, and capacity limitations on the agents are not exceeded. Two versions of the bottleneck generalized problem (BGAP) are defined. The first of these is called the Task BGAP and has as its objective the minimization of the maximum of the costs of the assignments that are made. The second version is referred to as the Agent BGAP and has as its objective the minimization of the maximum of the total costs assigned to each agent.

Applications of the BGAP arise in machine loading (scheduling) problems and also in facility location problems. For example, while most private sector facility location models involve a minisum objective function, many public sectors models (including the location of emergency service facilities) more appropriately deserve a minimax objective function. Since many facility location problems can be modelled as generalized assignment problems, the corresponding problem has the form of a BGAP.

The Task BGAP is appropriate when modelling an application for which the cost of any individual task assigned to an agent is deemed to be critical. For example, in the minimax version of the p-median problem it is desired to minimize the maximum response time to any of the demand points. This problem can be modelled as a Task BGAP. An application of the Agent BGAP arises in the machine loading problem. Here the costs typically represent job processing times on each of the machines. If the jobs assigned to different machines are performed in parallel, and if the jobs assigned to each machine are performed sequentially, then the time when all jobs are completed (the makespan) equals the longest time assigned to any of the machines. In this situation minimizing makespan involves finding a feasible assignment of agents to jobs such that the maximum agent-time is minimized. The problem can be modelled as an Agent BGAP.

Both the Task and Agent BGAP are formulated as mathematical programming problems. The structures of both problems are examined. Applications are discussed. Heuristic procedures as well as exact algorithms for solving both problems are presented. Finally, examples and computational results are given.

WEIGHTED MATCHING IN CHESS TOURNAMENTS

by

Snjólfur Olafsson
Science Institute
University of Iceland, Iceland

In many chess tournaments the number of players is much larger than
the number of rounds to be played. In such tournaments the pairing
for a round depends on the results in earlier rounds. Many different
systems have been constructed for these pairings. Two commonly used
systems are the Monrad system and the Swiss system though there exist
a number of different variants of these two systems.

In all these pairing systems there are two main goals: (i) players
with equal scores should play together; (ii) each player should
alternately play white and black. The first goal is the more
important and in the simpler systems, very little weight is put on
goal (ii). Usually there are other goals, working-rules, and
restrictions. One restriction in all these systems is that no player
may face the same opponent more than once.

Sometimes the pairing is very difficult. When a complicated system
that puts much weight on goal (ii) is used, it may take two or three
men 2-4 hours to achieve an acceptable pairing in a tournament with
between 50 and 100 players. Even so, they can expect to hear some
complaints from the players. Thus it is worth a great deal to have an
automatic way to do this pairing.

In Iceland, in a cooperation between a software house, the University
of Iceland and people involved in the management of chess tournaments,
a computer software for personal computers has been developed. This
software takes care of much of the routine work that the tournament
director and assistents need to do, not only the actual pairing, but
also, for example, writing tables showing the standing after each
round, and writing letters to FIDE concerning the ELO rates.

This paper discusses an algorithm for the pairing process and how weighted matching algorithm is used to solve this problem by converting the pairing rules into penalty points in such a way that the pairing with the least total number of penalty points is the pairing that best follows the rules. Somewhat simplified, it can be said that the solution algorithm (for one round) is a loop with three steps in each iteration:

(i)   Choose a group of players to be paired; stop if none is left. (The group is often a score group).

(ii) Define penalty points (weights) for each pair that has not played before.

(iii) Find a maximum cardinality matching with as few total penalty points as possible.

It is in steps (i) and (ii) that the transformation of the rules into an algorithm takes place. The number of penalty points is a sum of several factors containing some parameters. It is easy to adjust the algorithm to other variants of the Swiss system by giving these parameters new values.

In step (iii) a maximum weight matching problem on a graph is solved by using Edmonds blossoming algorithm.

The pairing system that we have used in this work is the Swiss system as approved by FIDE in 1985.

Keywords: Scheduling application, chess tournament, combinatorial optimization, weighted matching.

DECENTRALIZED OPTIMIZATION FOR STRUCTURED LINEAR PROGRAMMING MODELS

by

Cornelis van de Panne
Department of Economics
University of Calgary, Canada

This paper deals with methods for solving linear programming models with a primal or dual block-angular structure. A more general exposition of these methods can be found in [1]. For a primal block-angular structure the method is called the dual basis decomposition method.

A problem of that tructure is formulated as:

Maximize

$$f = \sum_{(k=0,1,..q)} p^k . x^k$$

subject to

$$\sum_{(k=0,1,..q)} A_k x^k = a, \quad x^0 > 0,$$

$$B_k x^k = b^k, \quad x^k > 0, \quad k=1,..,k.$$

Consider the independent subproblems:

Maximize

$$f_k = (p^k. - u´A_k)x^k \quad s.t. \quad B_k x^k = b^k, \ x^k >= 0.$$

and the principal problem

$$f_p = p^0. x^0 \quad s.t. \quad A_0 x^0 = a - \sum_{(k=1,..q)} A_k x^k, \ x^0 > = 0.$$

u contains the dual variables of the principal problem constraints.

The starting point is an optimal solution to each of the independent subproblems with u=0, and an optimal but not necessarily feasible solution to the principal problem. If this optimal solution to the principal problem is also feasible, the optimal solution to the entire problem is obtained. If not, an infeasibility of the principal problem is selected to be eliminated in the coming major iteration. This elimination is achieved by parametrically increasing this variable towards 0 in the entire problem.

Such a parametric increase is equivalent to a parametric increase in the principal problem and in each of the independent subproblems. If the first critical value found in the principal problems is u_p, the parametric solutions of the subproblems only have to be generated for critical values less than u_p. All critical values are then ordered according to increasing values, together with the corresponding accumulated changes of the infeasible variable of the principal problem. If these accumulated values do not exceed the infeasibility, all parametric steps of the subproblems are implemented, together with a pivot in the principal problem in the infeasible row. If the resulting principal problem solution is feasible, the optimal solution to the entire problem is found. If not, an infeasibility is chosen and another major iteration results.

If a parametric change in an independent subproblem eliminates the infeasibility before u_p is reached, all previous critical values are implemented, the corresponding independent subproblem is included in the principal problem, and a pivot in the included subproblem in the infeasible row eliminates the infeasibility. Then another infeasibility in the augmented principal problem is chosen for elimination in another major iteration. If the solution of the principal problem is feasible, the optimal solution to the entire problem must have been found, as optimality of all problems and feasibility of the independent subproblems is maintained by the parametric steps.

Subproblems that are independent are characterized by having the same number of basic variables and constraints, whereas subproblems included in the principal problem will have more basic variables than constraints. It can be shown that the number of subproblems included in the principal problem cannot exceed the number of common constraints. Subproblems may become independent again if a pivot in the principal problem makes the number of basic variables and constraints equal again.

A computer program has been written in APL for the dual basis decomposition method. It has been applied to randomly generated problems and the results have been compared with similarly programmed versions of simplex and dual methods. The results have been encouraging. The following table gives a comparison of CPU-times for the dual basis decomposition method and the dual method. There were 10 common constraints and the subproblems had matrices of 10x20. The number of subproblems varied from 1 to 10.

Number of Subproblems	Dual Basis Dec. Method	Dual Method
1	19.06	13.02
2	42.20	47.73
3	65.52	84.90
4	74.65	167.05
5	126.13	265.28
6	182.87	416.15
7	186.17	883.33
8	272.25	1482.13
9	384.25	>>
10	349.25	>>

Reference

[1] C. van de Panne, "Local Decomposition Methods for Linear Programming". European Journal of Operational Research Vol. 28 (1987), pp. 369-381.

TRANSPORT PLANNING IN AIRBORNE OPERATIONS

by

Adir Pridor
CIS Department
University of Delaware, USA

One of the most complex problems in military planning is that of transport in tactical airborne operations. It differs from transportation problems in several important aspects: it is time-dependent; it is multi-commodity with interrelations among commodities; it is capacity-constrained in a complicated manner; and, above all; it incorporates some severe military-tactical constraints. Some of these features can be found in scheduling problems, but even when viewed this way the planning process is very complex.

The main elements of an airborne operation are the following:
(1) Quantities of the various kinds of items that have to be transported, e.g. troops, vehicles of various types (jeeps, armored personnel carriers, tanks etc.), guns, air-defense launchers, ammunition and other supplied, and - possibly - more advanced items, like attack helicopters.
(2) Quantities and capacities of transport aircraft.
(3) Air-corridors and their capacities.
(4) Logistic data, including technical reliability.
(5) Attrition data.
(6) Tactical requirements concerning force consistency at the landing zone in various times.

This study suggests a model for this planning problem. By analysing the constraints it is shown that under quite general stipulations the problem can be formulated in terms of linear programming. The objective function can be set so that it refers both to the total elapsed time and to the number of transport aircraft needed for the operation. The model enables the military planner to deal with his problem using terms and expressions taken from the ordinary relevant tact cal-operational environment, and yet make good use of the strength and reliability of LP to get a solution fast. Post-optimal analysis, as interpreted in the model, provides the military planner with a convenient decision support tool useful when adjustments and modifications to the plan are considered.

The model has been applied to some real-world problems. One example is concerned with a divisional airborne operation in a framework similar to the U.S. Rapid Deployment Force. Problem formulation is based on tables and data from unclassified publications of the U.S. Army. The plan proposed by the model takes less time and makes more efficient use of transport aircraft than plans obtained by other available techniques.

A HIERARCHICAL APPROACH TO THE PLACEMENT PROBLEM

by

Maria Rosália Dinis Rodrigues
Departamento de Matemática
Universidade de Coimbra, Portugal

This paper concerns the problem of placing components on a circuit board, in such a way that a predefined set of objectives is optimized. The layout of integrated circuits motivated the use of hierarchical techniques as a means of dealing with increasing circuit complexity. A fully hierarchical approach to the placement problem is here described, based on a tree structure which embodies an adopted set of circuit properties and objective functions. The tree structure represents the relative positions of a hierarchy of blocks of components on the board surface.

Three objectives are here selected: minimal total wirelength and even distribution of both connections and components on the board area. A graph theoretical study is described, which investigates the correspondence between those objective and the structure of a binary tree.

A placement method is then described, consisting of two main steps: the building of the tree structure representing the circuit hierarchy and its subsequent embedding on different board environments. A set of algorithms is indicated, for the tree-building and tree-mapping on two basic types of boards; regularly structured and continuos plane.

OPTIMISATION OF OIL DEPLETION STRATEGY TO SATISFY LONG-TERM OBJECTIVES

by

R.V. Simons
SCICON LTD., England

The problem of determining how fast to extract oil from oil fields is a complicated one involving reservoir engineering, economic and political factors. Reservoir engineering simulations can be used to determine good ways of extracting the oil but they cannot do more than evaluate the economic consequences. By contrast, conventional MP models are unable to represent the behaviour of the oil reservoirs sufficiently accurately to be credible. A technique is presented for combining the accuracy of reservoir engineering simulations and the optimisation of the MP approach within the framework of a MP model.

Reservoir simulation studies are used to define a small number of profiles of how the reservoir should be exploited. These define the costs and flow of oil against time. The MP model is allowed to delay the start of a profile and, once it has started, to slow it down by taking less oil than is specified at that point on the curve of oil production capability against cumulative oil produced, which is regarded as invariant. An economic or political objective is incorporated and the model seeks to optimise this by selecting for each field a profile and the manner in which one proceeds along the profile.

The model is formulated as a mixed integer programming matrix with special ordered sets of type 2. Additional constraints are incorporated which define facets of the integer-feasible polytope and these have the effect of reducing run-times substantially.

APPROACHES TO OPTIMIZING FUEL CONSUMPTION IN CARS

by

Emilio Spedicato

Dipartimento di Matematica, Statistica, Informatica ed Applicazioni

Istituto Universitario di Bergamo, Italy

One of the most important problems which the automotive industry faces
is that of reducing fuel consumption, in view of the increasing cost
of energy and the need to save oil for more sophisticated uses, while
keeping pollutants emissions as low as possible, in view of the
resulting health and environmental problems. Reduction of fuel
consumption can be obtained working on different features of the car,
including type of engine, aerodynamical design, used materials and
optimal operations of the engine. In this work we shall be concerned
with the last approach, which can result in a 5-20% saving on fuel,
depending on which parameters are optimized. Mathematically, the
problem can be formulated as follows: let $s(t)$ be the value of a
vector of state parameters (typically speed and power) at time t; then
one has to choose optimal values of control parameters $u(t)$, typically
spark advance angle, air-fuel ratio and possibly transmission ratio,
in such a way that the total fuel consumption on a certain cycle (EPA
or European cycle) be minimized subject to constraints, of legal type,
on the total amount of certain emitted pollutants (typically CO, NO_x,
HC) during the cycle and of course to drivebility constraints. The
resulting problem is formally an optimal control problem, which is in
practice discretized and thus reduced to a mathematical programming
problem, of the nonlinear type and of moderately large dimensions (the
number of independent variables is in practice about fifty, with about
twice that number of constraints).

One of the main difficulties in actually solving the above problem is
that the form of the functions describing the rate of fuel consumption
in terms of $s(t)$ and $u(t)$ is not available analytically and the same
applies for the function describing the rate of pollutants emissions.
We review the different approaches taken in the literature to estimate
such functions, in particular the approach based on two-dimensional

splines which has been succesfully adopted by us. Then we describe the various techniques considered in the literature for actually solving the resulting mathematical programming problem, including simplifications to reduce its dimensionality. We describe the approach that is underway in Bergamo and which also allows sensitivity analysis. Finally we conclude with some considerations on the effect of car aging on the optimal control parameters and some ideas on how to update both the model and the optimal parameters.

EQUILIBRIUM DEFINITIONS IN SIMULATED ANNEALING:
A COMPUTATIONAL EXPERIMENT

by

Karl Spälti
Universität Bern, Switzerland

Recently simulated annealing, a probabilistic heuristic motivated by an analogy between statistical mechanics and combinatorial optimization, has been used to obtain good solutions to various combinatorial problems. The method minimizes a function f(s) whose arguments are the elements of a finite set S of feasible configurations (or solutions when applied to a combinatorial optimization problem) by a sequence of actions called local exchanges. All exchanges which improve f are accepted where as exchanges which increase f are accepted with a probability depending on a parameter called the temperature. At a given temperature level, a sequence of local exchanges, the so called inner loop iterations, are executed until the system reaches an equilibrium. The number of temperature levels and the way the temperature is lowered is called the annealing schedule.

When applying this procedure to a travelling salesman problem with n cities, the set of feasible configurations is the set of all 0.5*(n-1)! feasible tours. Several types of local exchanges, i.e. neighbourhood definitions, have been investigated in the literature [4]. For our purposes, we implemented the simple Lin neighbourhood and all computations were done with an identical annealing schedule.

To our knowledge, the only attempt to explicitly define an equilibrium in a simulated annealing heuristic has been made by Golden and Skiscim [2] who define an _epoch_ to be a specific number of inner loop iterations. At each temperature stage, after an epoch the final tour length is saved and compared with the tour lengths saved at the end of the previous epochs. If the tour length from the most recent epoch is sufficiently close to any previously saved tour length the system is defined to be in an equilibrium. For our purposes, we call a specific number of _accepted_ tours an epoch. Instead of comparing the length of the last tour of an epoch with the length of the last tours of all previous epochs, we compare the average tour length of an epoch with the average tour length of the previous one. By comparing the averages rather than the length of the final tours of the epochs we attempt to get a smoother representation of the changes in the objective function.

For small to medium size travelling salesman problems with up to 400 cities, our computational experiment with simulated annealing

procedures made it obvious that it is worth investigating various
equilibrium definitions for simulated annealing. The study has shown
the execution time of the heuristic using an equilibrium definition to
be significantly shorter than the one of a heuristic without an
explicit equilibrium definition without drastically worsening the
solution quality. Clearly, the trade off between computing time and
solution quality has to be judged in the light of a possible
application of the problem type, i.e. the travelling salesman problem
in our case.

References

1) V. Cerny, "Thermodynamical Approach to the Travelling
Salesman Problem: An Efficient Simulation Algorithm", JOURNAL
OF OPTIMIZATION THEORY AND APPLICATIONS Vol 45, 41-51 (1985).

2) B.L. Golden and C.C. Skiscim, "Using Simulated Annealing to
Solve Routing and Location Problems", NAVAL RESEARCH
LOGISTICS QUARTERLY, Vol 33, 261-279 (1986).

3) S. Kirkpatrick, C.D. Gelatt Jr. and M.P. Vecchi,
"Optimization by Simulated Annealing", SCIENCE, Vol 220, 671-
680 (1983).

4) Y. Rossier, M. Troyon and Th.M. Liebling, "Probabilistic
Exchange Algorithms and Euclidian Travelling Salesman
Problems", OR SPEKTRUM, Vol 8, 151-164 (9186).

Keywords:

Simulated Annealing * Travelling Salesman Problem * Heuristic

DECOMPOSITION IN INTEGER PROGRAMMING

by

Jørgen Tind
Department of Operations Research
University of Aarhus, Denmark

The objective is to discuss the construction of models for multilevel structures with the additional inclusion of integrality constraints.

Many multilevel structures without integrality constraints have been formulated in models and studied extensively in the framework of linear programming. A major step for the analysis of such models was taken through the development of the Dantzig-Wolfe decomposition procedure. Since, this has had a profound influence on the construction of models and algorithms for hierarchical planning problems in economics and management. See e.g. [1].

We shall here study the development of a similar decomposition procedure in the framework of integer programming in order to construct models for hierarchical planning systems where integrality requirements are present.

As usual, the procedure iterates in steps between a central level and one or more sublevels, each with its own set of constraints. At the central level feasible solutions are generated, based on solutions suggested from the sublevels. This step is performed via the solution of a special mixed integer programming problem. A convex polyhedral function in the space of constraints at the central level is generated simultaneously.

It is shown how this function can be obtained regardless of the choice among any of the main solution methods using branch and bound or cutting plane techniques.

Also as usual, the sublevels generate solutions in their sets of constraints. This is done by a linear programming routine, where the function, generated at the central level, is incorporated.

Finite convergence towards optimality is proven for the entire procedure.

Integer programming formulations of multilevel-planning problems can often be decomposed in alternative ways. This is finally discussed in relation to the modelling and algorithmic potential of the various alternatives.

The presentation is based on joint work with Søren Holm [2].

[1] Dirickx, Y.M.I. and L.P. Jennergren: "Systems Analysis by Multi-level Methods: With Applications to Economics and Management", Wiley, 1979.

[2] Holm, S. and J. Tind, "A Unified Approach for Price Directive Decomposition Procedures in Integer Programming". Publication no 87/2, Department of Operations Research, University of Aarhus, Denmark. (To appear in Discrete Applied Mathematics).

COMPUTATIONAL GEOMETRY AND LOW DIMENSIONAL LINEAR PROGRAMS

by

Henry Wolkowicz
Dept.of Combinatorics and Optimization
University of Waterloo, Canada
&
Adi Ben-Israel
Department of Mathematical Sciences
University of Delaware, USA

Many areas (robotics, pattern recognition, computer graphics etc.) offer problems requiring efficient computation of elementary planar geometric objects, e.g. the convex hull of a set, the intersection of a line and a convex polytope, the distance between two line segments. Computational geometry, applying computer science methodology to such geometrical problems (see e.g. [3], [4]), has provided important algorithms and complexity results, typically under specific assumptions on the data structure.

At the same time, there is need for practical algorithms with good expected performance, and fewer restrictions on the data structure. The approach chosen here is to model such geometric problems by linear programs $\max\{c^T x : Ax \leq b\}$, where A is m x n and n=2 or 3. The number of constraints m is typically large, e.g. when a convex set is approximated by a polytope.

For linear programs with small or fixed dimensions, the results in [1] and [2], see also [5], guarantee algorithms with linear time.

We propose here an algorithm for 2-dimensional linear programs which, at each iteration, bisects the feasible set, discarding about half of its constraints (and volume). We apply this algorithm to the problem of computing a common tangent to two disjoint polytopes (a step in computing the convex hull) and report numerical experience.

We discuss an extension of this idea to dimensions ≥ 3, and propose such an algorithm for linear programming, consisting of super-iterations (at each, 'half' of the feasible set is discarded, and at least one constraint is made redundant), and their iterations (where dimension changes possibly occur). The algorithm is unaffected by degeneracy.

REFERENCES

1) M.E.Dyer, Linear time algorithms for two- and three- variable linear programs, SIAM J. Computing 13 (1984), 31-45.

2) N. Megiddo, Linear programming in linear time when the dimension is fixed, J. Assoc. Comput. Mach. 31 (1984), 114-127.

3) K. Mehlhorn, Data Structures and Algorithms 3: Multi-dimensional Searching and Computational Geometry, Springer, Berlin, 1984.

4) F.P. Preparata and M.I. Shamos, Computational Geometry - An Introduction, Springer, New York, 1985.

5) A. Schrijver, Theory of Linear and Integer Programming, J. Wiley, New York, 1986.

LIST OF ADDRESSES

Prof. Carsten Aamand
Instituttet for Kemiindustri
DtH Bygning 227
DK-2800 Lyngby
DENMARK

Prof. Jaime Barcelo
Dept. d'Investigacio Operativa
Facultat d'Informatica de Barcelona
Pau Gargallo, 5
08028 Barcelona
SPAIN

Prof. Paulo Barcia
Faculdade de Economia
Universidade Nova de Lisboa
Campo Grande, 185
1700 Lisboa
PORTUGAL

Prof. Omer S. Benli
Dept. of Industrial Engineering
Bilkent University
P.O.Box 8, Maltepe
TR-06572 Maltepe, Ankara
TURKEY

Dr. Iain Buchanan
Dept. of Computer Science
Livingstone Tower
University of Strathclyde
Glasgow, GI 1XH
SCOTLAND

Prof. J.D. Coelho
Faculdade de Economia
Universidade Nova de Lisboa
Campo Grande, 185
1700 Lisboa
PORTUGAL

Prof. Marshall L. Fisher
Decisions Sciences Dept.
The Wharton School
University of Pennsylvania
Philadelphia, PA 19101
U.S.A.

Prof. Sjur D. Flåm
Dept. of Economics
University of Bergen
P.O.Box 25
N-5027 BERGEN - UNIVERSITETET
NORWAY

Prof. Robert Fourer
Department of Industrial Engineering
Northwestern University
Evanston, IL 60201
U.S.A.

Prof. Gus Gassmann
School of Business Administration
Dalhousie University
6125 Coburg Road
Halifax, Nova Scotia
CANADA B3H 1Z5

Dr. David M. Gay
AT & T Bell Laboratories
600 Mountain Avenue
Murray Hill, NJ 07974
U.S.A.

Dr. Åsa Hallefjord
Chr. Michelsen Institute
Fantoftvegen 38
N-5036 FANTOFT
NORWAY

Prof. Patrick T. Harker
Department of Decision Sciences
The Wharton School
University of Pennsylvania
Philadelphia, PA 19104-6366
U.S.A.

Dr. Ellis Johnson
IBM Corporation
Thomas J. Watson Research Center
Rt. 134, P.O.B. 218
YORKTOWN HEIGHTS, N.Y. 10598
U.S.A.

Prof. Kurt Jörnsten
Chr. Michelsen Institute
Fantoftvegen 38
N-5036 FANTOFT
NORWAY

Prof. Alan King
Dept. of Mathematics
The University of British Columbia
Vancouver, B.C. V6T 1Y8
CANADA

Prof. Stein Krogdahl
Dept. of Informatics
University of Oslo
P.O.Box 1080 - Blindern
N-0316 OSLO 3
NORWAY

Prof. Robert R. Meyer
Computer Sciences Department
University of Wisconsin
Madison, WI 53706
U.S.A.

Prof. J.M. Mulvey
School of Engineering and Applied Science
Princeton University
Princeton, NJ 08544
U.S.A.

Prof. Walter Murray
Department of Operations Research
Stanford University
Standford, California 94305
U.S.A.

Prof. Narcis Nabona
Dept. of Operations Research
Facultat d'Informatica U.P.C.
Universitat Politecnica de Catalonya
08028 Barcelona
SPAIN

Prof. Anna Nagurney
Dept. of General Business & Finance
School of Management
University of Massachussetts
Amherst, Mass. 01003
U.S.A.

Prof. Robert M. Nauss
School of Business Administration
University of Missouri-St. Louis
8001 Natural Bridge Road
St. Louis, Missouri 63121
U.S.A.

Prof. Alan W. Neebe
The University of North Carolina at Chapel Hill
Carol Hall 012 A
Chapel Hill, N.C. 27514
U.S.A.

Prof. Snjolfur Olafsson
Science Institute
University of Iceland
Dunhaga 3
IS-107 Reykjavik
ICELAND

Prof. Cornelis van de Panne
Department of Economics
University of Calgary
Calgary, Alberta T2N 1N4
CANADA

Prof. Adir Pridor
CIS Department
University of Delaware
Newark, DE 19716
U.S.A.

Samuel M. Rankin, III
U.S.A. Air Force Office of Scientific Researc
Directorate of Mathematics
Bolling AFB
Washington, DC 20332-6448
U.S.A.

Prof. A.H.G. Rinnooy Kan
Erasmus University
Econometric Institute
P.O.Box 1738
3000 DR Rotterdam
THE NETHERLANDS

Prof. Alan G. Robinson
School of Management
University of Massachusetts at Amherst
Amherst, MA 01003
U.S.A.

Prof. Maria R. D. Rodrigues
Departamento de Matematica
Universidade de Coimbra
Apartado 3008
3000 Coimbra
PORTUGAL

Mr. Jarand Røynstrand
Elektrisitetsforsyningens
Forskningsinstitutt
N-7034 TRONDHEIM - NTH
NORWAY

Prof. R.V. Simons
SCICON LTD.
Wavendon Tower
Milton Keynes, MK17 8LX
ENGLAND

Dr. Karl Spälti
Betriebswirtschaftliches Institut
Abteilung für Operations Research
Universität Bern
Ch-3012 Bern
SWITZERLAND

Prof. Emilio Spedicato
Dipartimento Matematica
Universita di Bergamo
via Salvecchio 19
I-24100 Bergamo
ITALY

Prof. Sverre Storøy
Dept. of Informatics
University of Bergen
N-5027 BERGEN – UNIVERSITETET
NORWAY

Dr. Jan Telgen
Van Dien+Co. Organisatie
Churchilllaan 11
3527 GV Utrecht
THE NETHERLANDS

Prof. Jørgen Tind
Institut for Operationsanalyse
Aarhus Universitet
Bygning 530, Ny Munkegade
DK-8000 Aarhus C
DENMARK

Dr. Stein W. Wallace
Haugesund Maritime College
Skåregaten 103
N-5500 Haugesund
NORWAY

Prof. Roger J-B Wets
1015 Department of Mathematics
University of California
Davis, California 95616
U.S.A.

Prof. Henry Wolkowicz
Dept. of Combinatorics and Optimization
University of Waterloo
Waterloo, Ontario N2L 3G1
CANADA

Prof. Georges Zaccour
Marketing
Ecole des Hautes Etudes Commerciales
5255 Decelles
Montreal, Que H3TT 1V6
CANADA

NATO ASI Series F

NATO ASI Series F

NATO ASI Series F